Lynne Hamill and Amparo Lasen (Eds)

Mobile World

Past, Present and Future

 Springer

Lynne Hamill
Amparo Lasen
Digital World Research Centre, University of Surrey, Guildford, UK

Series Editors
Dan Diaper, PhD, MBCS
Professor of Systems Science and Engineering, School of Design, Engineering and Computing, Bournemouth University, Talbot Campus, Fern Barrow, Poole, Dorset BH12 5BB, UK

Colston Sanger
Shottersley Research Limited, Little Shottersley,
Farnham Lane, Haslemere, Surrey GU27 1HA, UK

British Library Cataloguing in Publication Data
A catalogue record for this book is available from the British Library

Library of Congress Control Number: 2005927603

CSCW ISSN: 1431-1496
ISBN-10: 1-85233-825-3   Printed on acid-free paper
ISBN-13: 978-1-85233-825-1

Printed in the United States of America   (GP/EB)

9 8 7 6 5 4 3 2 1

Springer Science+Business Media
springeronline.com

Computer Supported Cooperative Work

Books are to be returned on or before
the last date below.

Also in this series

Celia T. Romm and Fay Sudweeks (Eds)
Doing Business Electronically
3-540-76159-4

Fay Sudweeks and Celia T. Romm (Eds)
Doing Business on the Internet
1-85233-030-9

Elizabeth F. Churchill,
David N. Snowdon
and Alan J. Munro (Eds)
Collaborative Virtual
Environments
1-85233-244-1

Christine Steeples and
Chris Jones (Eds)
Networked Learning
1-85233-471-1

Barry Brown, Nicola Green and
Richard Harper (Eds)
Wireless World
1-85233-477-0

Reza Hazemi and Stephen Hailes (Eds)
The Digital University – Building a
Learning Community
1-85233-478-9

Elayne Coakes, Dianne Willis
and Steve Clark (Eds)
Knowledge Management in the
Socio Technical World
1-85233-441-X

Ralph Schroeder (Ed.)
The Social Life of Avatars
1-85233-461-4

J.H. Erik Andriessen
Working with Groupware
1-85233-603-X

Christopher Lueg and
Danyel Fisher (Eds)
From Usenet to CoWebs
1-85233-532-7

Kristina Höök, David Benyon and
Alan J. Munro (Eds)
Designing Information Spaces: The
Social Navigation Approach
1-85233-661-7

Bjørn Erik Munkvold
Implementing Collaboration
Technologies in Industry
1-85233-418-5

Andy Crabtree
Designing Collaborative Systems
1-85233-718-4

David N. Snowdon, Elizabeth F. Churchill
and Emmanuel Frécon (Eds)
Inhabited Information Spaces
1-85233-728-1

Related Title
Richard Harper (Ed.)
Inside the Smart Home
1-85233-688-9

A list of out of print titles is available at the end of the book

Contributors

Imar de Vries, Faculty of Arts, Utrecht University, Utrecht, The Netherlands

Lynne Hamill, Digital World Research Centre, University of Surrey, Guildford, UK

Richard Harper, Microsoft Research, Cambridge, UK

Amparo Lasen, Digital World Research Centre, University of Surrey, Guildford, UK

Donna J. Reid and Fraser J.M. Reid, University of Plymouth, Plymouth, UK

Philippe Rouchy, Blekinge Institute of Technology, Karlskrona, Sweden

James Stewart, Institute for the Study of Science, Technology and Innovation, University of Edinburgh, Edinburgh, UK

Alex S. Taylor, Microsoft Research, Cambridge, UK

Richard Tee, University of Amsterdam, Amsterdam, The Netherlands

Jane Vincent, Digital World Research Centre, University of Surrey, Guildford, UK

Jenny Waycott, Smart Internet Technologies Cooperative Research Centre, RMIT Business, Melbourne, Australia

Contents

Introduction: Digital Revolution – Mobile Revolution?

Lynne Hamill

This book is a sequel to Wireless World: Social and Interactional Aspects of the Mobile Age, by Brown, Green and Harper published in this Computer-Supported Cooperative Work (CSCW) series in 2002. This new volume is a collection of invited chapters, drawing heavily on an international conference held by the Digital World Research Centre (DWRC) at the University of Surrey in July 2003. The title of the conference was *The Mobile Revolution – A Retrospective:* Its Theme, *Lessons on Social Shaping.* The idea was to focus on what can be learnt from the adoption of mobile devices that could be applicable to other, new, digital technologies.

The Digital Revolution?

What is it that makes digital technologies different from what has gone before? There are three key features that can be used to characterise digital technology. The first is that it reduces information – text, sound, picture or films – to a common currency of noughts and ones. It is this characteristic which leads to the convergence of devices: why cameras are now in phones, for example. The second, leading on from the first, means that it is easy to store information; and increasing miniaturisation means that it is very cheap to store very large quantities of data. Think about a film for example: what used to be stored on half a dozen large reels of celluloid is now stored on a disk you can put in your pocket. Thirdly, it is easy to transmit and reproduce. Nothing is lost in transmission or reproduction. For example, home audio taping of music was very inferior to vinyl records, whereas compact disks (CDs) can be copied with no loss of quality. This simple fact is having enormous implications for the music industry. What this adds up to is a step change, a discontinuity. The problem is trying to understand what the implications are, for businesses, public services and for society in general.

As it is a step change rather than more of the same, it presents a very big problem. In the words of Schumpeter (1934), it is "economic development" in the sense of a qualitative change that creates radically new conditions and for which technology is the prime driver. He gave this example: however many stagecoaches are produced, representing economic growth, you do not get a railroad, which represents economic development.

And there are possibly bigger implications too. The Russian economist, Kondratieff (1935 (1926)) identified long economic cycles or waves that lasted 50–60 years and were characterised by 20 years or so of growth and general prosperity, a plateau of some 10 years followed by 20 years of low growth. One of the main causes of these waves is attributed to what he called "changes in technique". He explained, "Changes in the technique of production presume that (1) the relevant scientific–technical discoveries and inventions have been made and (2) it is economically possible to use them. It would be an obvious mistake to deny the creative element in scientific–technical discoveries and inventions. But from an objective viewpoint, a still greater error would occur if one believed that the direction and intensity of those discoveries and inventions were entirely accidental; it is much more probable that such direction and intensity are a function of the necessities of real life and of the preceding development of science and technique." He went on to argue that the inventions will not bring about a change in production methods if the "economic conditions favourable to their application are absent".

This sounds very much like "social shaping" even though the phrase was probably not used until 1985 (Mackenzie and Wajcman, 1998). It implies that technology is endogenous in society: that it is developed within society rather than somehow being imposed from outside. It is a multidisciplinary concept. While economics is a key driver in the process, sociology and psychology provide insights into how people's needs are generated and how companies come to develop and design certain products and services.

Now Schumpeter (1939) identified three long waves, each of which, he argued, was driven by major technological change:

- 1780s to 1842: the "industrial revolution";
- 1842 to 1897: the introduction of the railways;
- 1897 to 1939: cars, electric power and changes in the chemical industry.

On the basis of this chronology, it is possible to make a case for two more waves: the "electronic revolution" running from the 1940s to the 1990s, followed by the "digital revolution", starting around 1990.

The immediate post-war period was characterised by boom conditions coinciding with the arrival of solid-state electronics. The transistor was invented in 1947. Texas Instruments claim to have produced the first commercial transistor radio in 1954 (Texas Instruments, 2004) and "by 1959 almost half of the 10 million radios made and sold in the US were the portable transistor type" (University of San Diego, 2004). And in 1959 integrated circuits appeared; many transistors on a single silicon chip performing a dedicated function. The first microprocessor, general-purpose programmable digital chip, appeared in 1971. By 1974 a microprocessor was powering the Altair, the first widely available personal computer, selling for under $400; equivalent to about £1,700 in 2004 (Intel, 2004; Ceruzzi, 1998).

The last decade of the 20th century saw a flood of new digital technology entering the marketplace, the office and the home. The CD player was introduced in the late 1980s, and by 1992 "had revolutionised the way we listen to music" (Central Statistical Office, 1994). Arriving on the commercial market in the early 1990s, digital cameras have transformed consumer photography. The digital video disk (DVD) player is said to be the fastest growing consumer good ever: virtually unknown in homes in 2000, by 2003 they were to be found in nearly half of UK households (OFCOM, 2004). The arrival of the Internet has transformed some businesses and contributed to the enormous growth of computers in the home. Mobile phones started as business tools in the 1980s but by 2002–2003 were to be found in 70% of UK households (ONS, 2004).

And on the horizon, there is more to come. In Chapter 4, Alex Taylor and Jane Vincent refer to "technology-in-progress" and in Chapter 8, Richard Tee refers to Mobile Internet Services as "a work in progress". Throughout this book, people talk about third generation (3G). But in the engineering laboratories, they are working on what comes after that, which some call fourth generation (4G). It appears that the digital revolution has just begun.

But what are the implications? What will life be like when the whole of the country, or indeed a large part of the world, is one big wireless hotspot? While mobile phones mean that you no longer have to be in your office or your home to make a phone call, this will mean that you no longer have to be at your personal computer (PC) to access the Internet. There are implications for almost all aspects of life: transport, health, education, and entertainment.

Yet, forecasting is a hazardous game, albeit essential for business and public sector service providers. In macroeconomic forecasting, which attempts to answer quantitative questions such as how much the economy will grow next year, it is easy enough if next year is expected to continue on current trends, with all the main variables expected to be within a couple of percentage points of their current values. The key problem is foreseeing turning points or discontinuities: an unexpected oil price hike, for example, that will destabilise the economy.

If the introduction of new technology is regarded as a turning point or discontinuity, then it is easy to understand why it is so difficult to forecast its impact. As Carey and Elton (1996) point out, "The past century is littered with erroneous forecasts and predictions. Some have seriously underestimated demand; most have overestimated demand".

Overestimating demand can have enormous economic costs: not just the financial costs – the money lost – but also the "opportunity costs", the technology that might have been developed if time and effort had not been devoted to the failed product. For example, it is reported that American Telephone and Telegraph (AT&T) predicted that one million picture phones would be in use by 1980, and two million by 1985. But picture phones failed to take off at all (Carey and Elton, 1996).

But, occasionally, forecasts seriously underestimate. For example, it is reported that the founder of Digital Equipment Corp said in 1977: "There is no reason anyone would want a computer in their home". Yet by 2002–2003 half of UK homes had one (ONS, 2004). The mobile phone industry was subject to serious underestimation too. Demand for the phones themselves, it is said, was underestimated by an order of magnitude: hundreds of thousands in the UK rather than the millions who now have them. Also, as discussed by Alex Taylor and Jane Vincent, in Chapter 4, demand for certain services, in that case Short Messaging Services (SMS), was also seriously underestimated.

Of course, history never repeats itself exactly. As the adverts for stock market investments say "past performance is no guarantee of future performance." But past performance is all there is to base forecasts on, whether that experience is expressed in terms of mathematical equations based on large quantitative data sets or on qualitative ethnographic observation. The hope is that this book will provide some ideas that might contribute to reducing these expensive forecasting errors in the future.

The Mobile Revolution

The idea of person-to-person wireless communication is not new. In 1892, Sir William Crookes foresaw person-to-person wireless communications and in 1901 Prof. William Ayrton predicted: "a time when if a person wanted to call to a friend he knew not where, he would call in a loud, electromagnetic voice, heard by him who had the electromagnetic ear, silent to him who had it not. 'Where are you?' he would say. A small reply would come, 'I am at the bottom of a coal mine, or crossing the Andes, or in the middle of the Pacific.' Or, perhaps, in spite of all the callings, no reply would come, and the person would then know that his friend was dead. Let them think of what that meant, of the calling which went on every day from room to room of a house, and then think of that calling extending from pole to pole; not a noisy babble, but a call audible to him who wanted to hear and absolutely silent to him who did not."

By the 1920s, police forces in the USA were using radios to communicate. The first basic "Mobile Telephone Service" was launched in the USA in 1946. In 1955, the first was launched in Europe in Sweden, where by 1981, there were 20,000 mobile users (Agar, 2003). By 2004, the number of mobile phone users worldwide had reached 1.2 billion (GSM, 2004), representing an astounding one in six of the global population. Mobile phones, especially in Europe and the Far East, have become ubiquitous and their use is rapidly becoming an unnoticed and taken for granted aspect of everyday life. As mobiles have become more widespread, all sorts of issues have been and are being raised about the effect of this technology on people and society.

This book brings together the work of researchers in arts, economics, information science, psychology, science technology studies, and sociology. Geographically, it ranges across Europe and beyond to Japan. Most importantly, it deals with people and how they use new technology in their everyday lives, both at home and at work. It aims to examine the past for clues as to what may happen when new technology is introduced in the future. A key underlying theme is that people and their needs do not change, but that new technology changes the way that these needs are met. A better understanding of people's needs would provide a better idea of what technology people really want to use.

It also explores how social scientists can collaborate with designers and engineers in the development of new devices and uses. It is divided into three major sections: lessons from the past, present users, and how to study the future.

Part 1, which looks at lessons from the past, starts with Imar de Vries' Mobile Telephony: Realising the Dream of Ideal Communication? This chapter applies the concept of ideal communication to media in general, and to mobile telephony in particular. It argues that mobile telephony, as the most recent addition to our media spectrum, may seem to achieve communication Utopia. However, experience and expectations of this medium are surprisingly similar to those of older media, and it is argued that, therefore, mobile telephony is bound to fail to meet this high expectation, as did its predecessors.

In the second chapter, History Repeating? A Comparison of the Launch and Uses of Fixed and Mobile Phones, Amparo Lasen compares the history of society's adoption of the fixed-line telephone and the corresponding adoption of the mobile telephone. This comparison covers a time span of a century and yet, considering the differences of social contexts and technical devices, the knowledge of early practices, conflicts, fears and hopes about telephones can nevertheless help to understand the uses and social roles of mobile telephones. The interest of the comparison is, therefore, to give an insight into what happens when new services and new devices enter a marketplace.

In Chapter 3, Kids will be Kids: The Role of Mobiles in Teenage Life, Richard Harper and Lynne Hamill look at two aspects of life: mobile phone bills and social etiquette. Focusing on teenagers, and drawing on evidence from the 1960s and recent work on teenagers and mobiles, they advance the hypothesis that what teenagers do has not changed significantly with the arrival of mobiles, but the new technology has provided a new way for them and their parents to do what they have always done. In Chapter 4, Alex S. Taylor and Jane Vincent look at An SMS History, telling a tale of social shaping. They explain how technical, economic and social factors intertwined to produce a totally unforeseen demand for a service. Drawing on this experience, they look forward to what lessons might be drawn from this tale to apply to camera phones.

Part 2 focuses on the present, starting with Jane Vincent's Emotional Attachment to Mobile Phones: An Extraordinary Relationship. This explores people's relationship with their mobile phones and argues that it embodies an emotional attachment, and mobile phone use involves emotional behaviours. The chapter describes those behaviours, both in the way that people are observed using their mobiles and in the terms they use to describe their relationship with them. It offers evidence from recent empirical research of how attached people have become to their mobiles and it offers some explanations. Finally the chapter suggests some of the implications that these behaviours might have on the development of mobile communications and how these are different from that of any other information communication technology devices.

In Chapter 6, Textmates and Text Circles: Insights into the Social Ecology of SMS Text Messaging, Donna Reid and Fraser Reid look at the psychological impact of texting on social interaction amongst regular users and ask whether texting has the same effect as is reported for Internet use. They present here preliminary findings of a survey, with particular reference to measures of phone usage, patterns of communication, social anxiety and loneliness. Exploring the differences between those who prefer texting and those who prefer talking on their mobiles, it is found that "Texters" had a close-knit text circle, interconnecting with a close group of friends in perpetual SMS messaging contact. Rather than providing a stepping-stone to real-world relationships, texting provides a distinctive and alternative mode of social contact which continues to appeal to Texters even when textmates are already close friends meeting up on a regular face-to-face basis. Furthermore, the evidence suggests that texting does help to create relationships that may not have developed otherwise.

Next, in Appropriating Tools and Shaping Activities: The Use of PDAs in the Workplace, Jenny Waycott moves away from mobile phones and looks at the use of Personal Digital Assistants (PDAs) as general workplace tools. She reports what happened when staff from two very different organisations used PDAs for various purposes, including to support time management, to access and manage e-mails, and to record and store notes. She uses activity theory as a framework for understanding how new tools mediate activity. This provides a useful vocabulary for describing how mobile technologies can disrupt and change workplace activities: for example, by changing the relationship between elements of the activity, by modifying the processes by which the activity is carried out and by introducing and resolving contradictions.

Part 3 turns to ways of studying the future. Starting with Different Directions in the Mobile Internet: Analysing Mobile Internet Services in Japan and Europe, Richard Tee makes use of a Social Construction Of Technology (SCOT) approach to identify the underlying causes that have shaped the developments of Mobile Internet Services in Europe and Japan, respectively. The configuration of relationships between relevant social groups is identified as a key factor. In particular, the power of the mobile operator in Japan has had far-reaching consequences with regard to the way Mobile Internet Services can be introduced and controlled. While no actor in Europe has either the power or the will to coordinate Mobile Internet services as these have been introduced in Japan, the Mobile Internet has taken a different shape, characterised by protocols based on open standards as well as proprietary platforms.

In Chapter 9, Context Perspectives for Scenarios and Research Development in Mobile Systems, James Stewart examines the need for tools to link the work of engineers developing future wireless technologies to the vast amount of existing social science research, particularly qualitative, is imperative. To do this, a common language and set of frameworks is needed so the apparently rather different concerns of each group can be met. Stewart sets out a framework for developing scenarios by looking at the interaction between people, devices, location and application packages.

Finally in Chapter 10, Instant Messaging and Presence Services: Mobile Future, CSCW and Ethnography, Philippe Rouchy explores Unified Messaging Services (UMS) using ethnography. He explains the context in which UMS technology is developed and then demonstrates how evaluative ethnography can work as a tool of fruitful comparison between present and future, as a first proxy to the use of a yet-to-come technology. This he does by looking at the role of ethnography in CSCW and then drawing on four vignettes, designed to draw out specific features of the new technology. In doing this, he draws interesting comparisons between the concepts of users and consumers; and between usability and usefulness. He concludes that going from CSCW studies of distributed work in large organisations to CSCW studies of the use of mobile technology in private and individualised settings allows forthcoming technology to be assessed. In particular, evaluative ethnography can help to formulate socio-economic dimensions of users rather than technical design suggestions.

Acknowledgements

I would like to thank those without whom this book would not have been possible; in particular, past and present members of the Digital World Research Centre: Nigel Gilbert, Richard Harper, Alex Taylor, and Jane Vincent, with very special thanks to Amparo Lasen, who did much of the early work on this book, and to Caroline Martin, without whose administrative efforts the conference on which this collection is based would not have happened.

References

Agar J (2003) *Constant Touch*. Icon, Cambridge.

Ayrton WE (1901) *Electrical Review*, 29 June, p. 820. http://earlyradiohistory.us/1901ayrt.htm

Brown B, Green N, Harper R (eds) (2002) *Wireless World: Social and Interactional Aspects of the Mobile Age*. Springer, London.

Carey J, Elton M (1996) Forecasting the demand for new consumer services: challenges and alternatives. In: Dhokalia RR, Mundorf N, Dhokalia N (eds). *New Infotainment Technologies in the Home: Demand-Side Perspectives*. Lawrence Erlbaum Associates, New Jersey.

Central Statistical Office (1994) *Social Trends 24*, HMSO, London.

Crookes W (1892) Some possibilities of electricity. *Fortnightly Review*, 1 February pp. 174–176. Available from: http://earlyradiohistory.us/1892fort.htm

Ceruzzi P (1998) Inventing personal computing. In: MacKenzie D, Wajcman, J (eds). *The Social Shaping of Technology*. Open University Press, Buckingham and Philadelphia; pp. 64–86.

GSM World (2004) http://www.gsmworld.com

Intel (2004) http://www.intel.com/intel/intelis/museum/online/hist_micro/hof/index.htm

Kondratieff ND (1935) The long waves in economic life. *Review of Economic Statistics*, **17**(6): 105–115 (translated from (1926) Die langen Wellen der Konjunktur. *Archiv fur Sozialwissenschaft and Sozialpolitick*, **56**(3): 573–609).

MacKenzie D, Wajcman, J (eds) (1998) *The Social Shaping of Technology*. Open University Press, Buckingham and Philadelphia.

OFCOM (2004) The Communications Market 2004. http://www.offcom.org.uk

ONS (2004) Family Spending 2002–2003. TSO, London.

Schumpeter JA (1934) *The Theory of Economic Development*, first published by Harvard University Press (reference to 1983 edition).

Schumpeter JA (1939) *Business Cycles: A Theoretical, Historical and Statistical Analysis of the Capitalist Process*. McGraw-Hill, London.

Texas Instruments (2004) http://www.ti.com/corp/docs/company/history/radio.shtml

University of San Diego (2004) http://history.acusd.edu/gen/recording/transistor.html

Part 1

Lessons from the Past

1

Mobile Telephony: Realising the Dream of Ideal Communication?

Imar de Vries

1.1 Introduction

Every existing communication technology was once new and full of promise. Upon their introduction, numerous producers and consumers were euphoric about the potential for the immediate application and had fantastic visions of future use. Some of them even cherished the notion that all previous media would be rendered obsolete. Wireless telegraphy was seen as "the means to instantaneous free communication" (Flichy, 1995: 109); telephony seemed to promise banishment of distance, isolation and prejudice (Briggs, 1977: 45); radio would pave the way for contact with the dead (Sconce, 2000: 59–61) and television would transform its viewers into eyewitnesses of everything that went on in the world (Elsner *et al.*, 1994: 110).

Most of these expectations never really came to fruition, almost every medium has developed in a different way from what was foreseen and no medium so far has completely replaced all other media. Nevertheless, the process of praise for unprecedented opportunities remains conspicuous throughout media history. Today, this phase of media hype can unmistakably be recognised in the way we think and talk about mobile telephony, one of the most recent and widely used communication media we have come to know.

This observation of ever-repeating appraisal gives rise to the assumption that there is a prevalent idea of which role communication media should ideally play in our lives, an idea that not only influences the design process, development and actual social use of communication technology, but is also reshaped and reinforced by these steps. As John Durham Peters notes in *Speaking into the Air*: "Communication is a registry of modern longings. The term evokes a utopia where nothing is misunderstood, hearts are open and expression is uninhibited. ... each medium ... was an attempt to cover a human lack, to fill the gap between ourselves and the Gods" (Peters, 1999: 2, 219).

It seems that, each time mankind realises that no medium has yet fulfilled the utopian idea of ideal communication, it is urged to improve the technology it has used up to that point. Thus, media evolution can be viewed as the continuing search for an ideal medium, which in the end has to comply with all the demands that the idea of ideal communication imposes on its characteristics.

This provocative viewpoint needs two annotations. First, the concept of "ideal" is of course problematic, in the sense that its interpretation is an exceedingly personal matter. Throughout this chapter, "ideal communication" will be referred to, in line with Peters (1999) and Katz and Aakhus (2002), as perpetual contact, the fulfilment of "sharing one's mind with another"; in other words, as ubiquitous and pure communication without misunderstanding. Second, dystopian visions are as much part and parcel of the reception of media as utopian ones. To say that we expect nothing but good from new media technology would be very naive, considering everyone can think of examples in which technology was the cause of chaos, disaster or havoc. However, this chapter will focus mainly on the positive expectations that continue to exist, in spite of the fact that they are always rebutted by actual media experience.

Suggesting that all media so far have been intermediate versions of one final and ideal medium could imply there is a grand direction that media technology heads towards, independent of social, cultural, political, economic or other factors that exert influence on the actual process of technology design, development and use. Because such a straightforward approach would deny the unpredictable and erratic ways in which media evolve, Section 1.2 tries to establish a more elaborate model of the relation between technology and society. In Section 1.3 the extent to which "old" communication media (the telegraph, telephone, radio and television) have realised properties of the concept of ideal communication is discussed. This in turn serves as a background to Section 1.4 to determine how mobile telephony relates to media evolution and if it really is the best addition to our media spectrum yet. This chapter concludes in Section 1.5 with the notion that all media have evolved and found their place in society in different and unpredictable ways, but did so with the same drive of trying to realise ideal, angelic communication.

1.2 Technology and Society

In his study of how telephony became integrated into the lives of Americans in the first decades of the 20th century, Claude Fischer (1992) argues that there are three possible theoretical models to describe the relation between technology and society. First he discusses the *method of impact analysis*, which can best be understood as a "billiard ball model". Here, technology enters society from the outside, has a certain impact on certain elements in that society, which in turn have an effect on other elements and so forth until the force of the initial impact has ebbed. Fischer dismisses this model for being too deterministic and for not taking into account specific culturally determined uses of new technology. The "softer" version of this model, the so-called imprint–impact model, tries to overcome this criticism by explaining the impact as a process in which "(the) essence of technology transfers itself to its users," the sudden ring of a telephone for instance, would cause feelings of fear and anxiety in its users. Still, Fischer finds this model too deterministic, and argues that proof for the existence of such a transfer of "essence" is, at the very least, speculative (Fischer, 1992: 8–11).

The second model is that of *symptomatic approaches*, which sees technology not as a force that enters society from the outside, but rather as an expression of processes (understood as the Hegelian notion of Geist) within that society. Again, according to Fischer, this approach should be dismissed because it is very difficult to determine exactly what these underlying processes consist of, and one might be wrongly persuaded to extrapolate a "grand direction" from what could merely be short-term technological expressions. Ironically, Fischer illustrates this latter problem by referring to the video game and computer industry, for which a bright future was predicted in the mid-1980s, though, at the time Fischer wrote his book, this was never realised (*ibid.*: 12–16).

The third model he describes is *social constructivism*, which sees the ultimate application of certain technologies within society as the result of conflicts and negotiations between those parties that have an interest in the development of those technologies. Fischer argues that using this social constructivism model, one is best equipped to take into account the various ways in which users of technology have actually incorporated it into their lives. This method would also recognise how political, social, cultural and economic factors play a role in establishing the environment in which those conflicts and negotiations take place (*ibid.*: 16).

Interestingly, applying this third model can, just as with the second model, lead to unfolding the underlying processes within society that influence the development of technology. In his essay "New Technologies and Domestic Consumption", Eric Hirsch (1998) uses a social constructivist method to conclude that during the past 200 years, the relation between socio-cultural innovations (such as the advent of the nuclear family) on the one hand and technological innovations (such as radio or television) on the other, has strengthened (Hirsch, 1998: 159). Technological development would therefore primarily exist to maintain that relationship keeping consumers and producers of technology mutually dependent on each other and legitimising each other's existence.

Peters is clear in his philosophical judgement of the role media technology has in our social lives. He names distance and death as the two largest obstacles in our eternal quest for love, thereby constantly raising desire, and argues that we have always tried to find ways of overcoming those obstacles: "Eros seeks to span the miles, reach into the grave, and bridge all the chasms. It is the principle that seeks to transcend the limitations of our normal modes of contact with each other in word and in the flesh. New media, by smashing old barriers to intercourse, often enlarge Eros's empire and distort its traditional shape, hence they are often understood as sexy or perverse or both" (Peters, 1999: 137). Drawing on Augustine's theories of the sign, Peters describes the ultimate goal of this human longing as the ability to communicate like angels and states that these angels "haunt modern media, with their common ability to spirit voice, image, and word across vast distances without death or decay" (*ibid.*: 75).

Peters's observation, combined with Hirsch's use of the social constructivist method, gives rise to the idea that Fischer's symptomatic approaches model does not have to be entirely dismissed for being too problematic. Fischer's unfortunate example of the danger of extrapolating a grand direction even supports this idea. It is plausible to combine both social constructivism and symptomatic approaches, to come to a critical symptomatic theory of technology and society. Such a theory would take into account the social, cultural, political and economic factors that determine the actual individual use of technology on the one hand, but on the other hand also recognise the processes that have always been present during technological development, as a steady undercurrent, a media Geist influencing the way we think about, use, develop and approach communication technology.

As an example of how such a combined theory should work on both levels of understanding media evolution, let us look at Marshall McLuhan's prophecy of the global village. According to McLuhan (1964), media are to be considered extensions of the human body, enabling users to, for instance, hear or see farther, or to bridge distances faster than before. Often though, human fascination for progress and improvement turns out to be strong enough to let us forget about the negative consequences of these extensions. One of these negative consequences, McLuhan argued, was the diminished importance of oral culture, which was brought about by the invention of writing and the Gutenberg press, and which had steered humankind away from its authentic "tribal" self. Luckily for humankind, according to McLuhan, the return of this tribal culture would be established by yet another extension, namely electronic media. This technology would eliminate distance and time, and create a global village in which all inhabitants would have the opportunity to contact one another.

Looking at today's pervasion of our global society by electronic media, one might be inclined to say that McLuhan was right, and that media evolution indeed has a tendency to keep humans close to their tribal self (Levinson, 1997). However, this would oversimplify the situation and not take into account the way people actually use electronic media, or even whether they have access at all. As Castells wrote after extensively studying the network society, "While the media have become indeed globally interconnected, and programs and messages circulate in the global network, we are not living in a global village, but in customized cottages globally produced and locally distributed" (Castells, 1997: 341).

The world is undeniably more interconnected than it was 200 years ago, but not everyone is connected in the same way, or their connection for the same purposes. The critical symptomatic theory mentioned above recognises both the underlying force that keeps the balance of extensions of our sense organs intact, as well as the notion that actual media use is subject to the individual's social, cultural, political and economic environment.

To determine whether the idea of ideal communication, understood as the desire to eliminate distance (and therefore time), and remove all obstacles on our way to reaching Eros, can be considered as the vital media Geist that has influenced the way communication technology has evolved so far. It is necessary to look for signs in our media history that suggest the existence of a prevailing human desire to improve "imperfect" media. The process of taking "old" media and turning them into something "new" is dealt with in the next section.

1.3 Media Evolution

As discussed, the desire to eliminate distance and time can be said to exert great influence on the development of communication media. For, if it really does play a major role, the search for ideal communication would dictate that each new medium must bring us closer to utopia. The process of media evolution does seem to follow a regular pattern, as each new medium boasts to be superior to all previous media (Bolter and Grusin, 1999: 14–15). Radio was telephony without wires; television was radio with pictures. Taking the characteristics of old media and adding to or improving on them is called remediation. According to Bolter and Grusin, the existence of remediation can be ascribed to the desire for transparent immediacy, a longing for experiencing the mediated world without being conscious of the involvement of a medium. Just as in Peters's angelic communication, we strive for transparent immediacy in order to forget that we have a body, and that we must use technology to transcend distance and time.

Paradoxically, the transparent immediacy component of remediation is always accompanied by one of hypermediacy: every time a new medium distances itself from other media by promising a more immediate experience, "the promise of reform inevitably leads us to become aware of the new medium as a medium" (ibid.: 19). In other words, transparent immediacy is never fully realised and it never will be. Again, the tragic desire to improve media is prompted by recognising that earlier attempts to fulfil the ultimate goal have failed so far: "The mistake is to think that communications will solve the problems of communication, that better wiring will eliminate the ghosts" (Peters, 1999: 9). This self-consuming snake is visible throughout media history. The next subsections take a look at how the telegraph, telephone, radio and television have each promised to take us closer to ideal communication.

1.3.1 The Telegraph

The telegraph was the first medium that freed communication over distance from physical transport of information. Before the end of the 18th century, people needed pigeons, boats or other messengers; with the advent of the telegraph the message lost its physical carrier. In the first half of the 19th century, Morse's electrical telegraph speeded up this process: "The telegraph ... fits precisely into the lineage of Augustine, the angels, and Mesmer: communication without embodiment, contact achieved by the sharing of spiritual (electrical) fluids" (Peters, 1999: 139). To many, this separation led to the belief that "electricity [could] mingle souls" and that the telegraph provided an earthly form of the way angels communicated (*ibid.*: 94). Despite inevitable start-up difficulties, the distances that could be bridged looked set to be endless, possibly not even bound to the physical world. It is not surprising that, immediately after the introduction of the telegraph, rumours began to circulate that one could "telegraphically" contact the dead. This Spiritual Movement, which had its high point in the 1870s, took the bodiless form of telegraphic communication as a conceptual model for "a land without material substance, an always unseen origin point of transmission for disembodied souls in an electromagnetic utopia" (Sconce, 2000: 57).

The telegraph was also the first medium that defined and established the basic characteristics of a telecommunication network in three ways. Firstly, it used a standard "language" to communicate, making transmission seamless and fast. Secondly, the network became permanently established and could grow by adding connections. Thirdly, the network was operated and overseen by specialised technical management, benefiting maintenance (Flichy, 1995: 32). All these characteristics are still visible in today's modern communication networks.

The utopian idea of everyone being able to communicate with anyone else was not realised, however, for the network was still small and could only make limited point-to-point contacts. Moreover, direct conversation was almost impossible because of the time it took to encode, send and decipher messages. The common man could hardly come into contact with the medium, for the telegraph was mainly used by the states, big companies, businessmen or rich civilians. This was because of state monopolies, the difficulty of operating the machinery (one had to learn Morse code), the notion that such a new medium was to be used for serious business only and cost.

1.3.2 The Telephone

This all changed dramatically with the introduction of the telephone. First, it took away the necessity of using an intermediary who knew Morse code. This enhanced the process of communicating via electricity to such an extent that this breakthrough was literally accessible to everyone. Once sending and receiving messages had become less troublesome and less formal, the nature of these messages also became more informal. While the leaders of the telephone industry first saw this frivolous usage as undesirable (see Fischer, 1992: 81–83), the characteristics of the telephone actually corresponded perfectly with the 19th century's expectation of being able to use electricity to communicate like angels, in which there was no need to differentiate between serious and non-serious messages. After the initial expense and costs of use had dropped, the proportion of people using the telephone for social purposes rose to great heights, indicating that its initial business-like character was merely inherited from the telegraph and was not specifically related to the medium itself.

A second effect of the telephone was the creation of a permanent global network. The prediction, introduced by the telegraph, that electricity would enable anyone to contact anyone else, quickly became a reality for those who bought themselves a place in the continuously growing network. This augmentation of nodes did not necessarily result in the formation of new social networks and people did not suddenly want to reach others purely because they were far away, but the telephone became a great tool for maintaining existent networks, even if these had expanded greatly under the influence of, for instance, urbanisation. According to Colin Cherry (1977), it is precisely this property of the telephone – to act as an information exchanging node in a network, that makes the impact on our experience of distance and time so immense: "The exchange principle led rapidly to the creation of networks, covering whole countries and, since World War II, interconnecting the continents. Anybody, without special training, can move about the geographic areas covered by the network and yet appear to another person on the network to be stationary" (Cherry, 1977: 114).

The idea that the telephone improves communication, therefore bringing us closer to the utopian angelic state, vibrates through its history. The installation of a secret "Hot Line" between the White House and the Kremlin on 30 August 1963, just months after the nearly disastrous Cuba crisis was finally brought to an end, is a clear example of the expectation that quicker and more direct communication would eventually lead to less misunderstanding and even the possibility of world peace (CNN Spotlight, 2000). Alas, as with the telegraph, this dream was soon found to be infeasible. The absence of visual cues meant that the risk of miscommunication was greater than with face-to-face exchanges; moreover people could only be reached if they were in the vicinity of the apparatus. On top of that, the telephone network required a lot of wires, and could only establish point-to-point contacts.

1.3.3 The Radio

The next step in media evolution, radio, promised to help solve most of the problems of the telephone. It was the first medium to realise the idea of broadcasting, by freeing information transfer from the wire and having it widely dispersed through the use of radio waves. The properties of the radio wave, being able to reach any point unhindered, having no specific destination and potentially being picked up by anyone with a receiver were initially seen as serious defects (Douglas, 1987). Radio came into existence at a time when the telegraph and telephone dominated the way people thought of communication media, so it seemed logical to think of radio as a means to perfect point-to-point contacts. However, as radio amateurs (and at a later stage the radio industry itself) soon discovered, these "defects" opened up a whole new form of communication. People started to broadcast music, news and plays, and found an incredible feeling of power in this act, "seeing the wireless as a utopian form of communication that would bring the nation closer together in a truly democratic fashion" (Spigel, 1992: 26). Radio programmes became less formal and tried to give their listeners a feeling of proximity to the broadcaster and to other listeners, to bridge "obvious gaps of distance, disembodiment and dissemination" (Peters, 1999: 211). The resulting communication model was aimed at providing entertainment, information and, above all, a feeling of "we-ness" (a sense of cohesiveness in a group, see Lewin, 1951). This idea of unity was all the more present because millions of listeners lived through the same radio schedule: the broadcasting pattern brought the experience of time directly into the living room (Moores, 1988: 35–38).

The wonder of hearing at a distance via invisible and all-penetrating radio waves fuelled the dream of crossing immense distances, bringing angelic communication closer than ever. To many, it did not matter what they were listening to, the sheer possibility of reaching out and touching someone was enough to spend countless nights "DX-fishing", searching the radio spectrum for distant voices. The prospect of stumbling upon alien signals sparked a true Mars-mania, for "contacting Mars would be the high priority of the new radio future and the ultimate 'catch' of DX-fishing" (Sconce, 2000: 102). Still heavily influenced by the Spiritualist Movement, some even went as far as believing radio could overcome separation by death. Lodge, Edison and Marconi all worked on apparatuses for contacting the dead (*ibid.*: 60–61). Compared to these prospects, telepathic contact seemed all the more likely to come into existence. Sir William Crookes's article "Some Possibilities of Electricity", written in 1892, assumed it possible to communicate without technology, thanks to "brain waves". At a time when radio waves had just been discovered, there seemed no reason to believe Crookes's brain waves were any different from the vibrations radio made in the ether. As long as we could use "proper tuning", it was argued, ideal or telepathic communication would become a reality (Peters, 1999: 104).

But the realisation of the dream of ideal communication was not to be. The free character of radio quickly disappeared with the rise of radio networks, which, with help from state organisations such as the Federal Communications Commission in America, took control of the available radio frequencies.

While radio practice geared towards entertainment and reporting on news events, looking for contact with the dead or with aliens became the domain of less seriously taken pseudo-science. Listeners slowly became aware of what it was like to be part of "an invisible scattered audience", and to know of (horrible) events far away, while being physically isolated from them (Sconce, 2000: 62). Above all, the balance of communication now seemed to lean heavily on oral and audio contact alone.

1.3.4 The Television

Television corrected the imbalance found in the way earlier media favoured the ears. It extended the basic property of the telegraph, telephone and radio, that of establishing audio contact between two or more physically separated points by adding moving images. For the first time in history, it became possible to see what was happening at the same time somewhere else, fulfilling a desire for visual liveness and simultaneity that, up to the introduction of television, had mostly (but not really) been satisfied by early non-fiction films (Uricchio, 2000). John Logie Baird's claim in 1926 that the videophone had finally been realised, after it had been prophesied decades before, initiated a considerable step in media history. Television was simply welcomed as the most recent medium that would subjugate space and time to our command: "Given its ability to bring 'another world' into the home, it is not surprising that television was often figured as the ultim-ate expression of progress in utopian statements concerning 'man's' ability to conquer and to domesticate space" (Spigel, 1992: 102).

McLuhan's idea of a global village began to take shape: with the help of electronic media, we would be able to remove the obstacles of time and space from our quest for presence and contact, wherever we might be, restoring the tribal community on a global scale. The "desire for physical participation with spatial proximity", which could only be fulfilled in theatres, cinemas or other public places, could now also be satisfied by television, by delivering a "surrogate for being present" (Elsner *et al.*, 1994: 136).

As with radio, a "fantasy … of imaginary unity with 'absent' others" origin-ated from the use of television, stemming from the idea that millions of people watched the same programme, or were aware of the same events happening in the world (Spigel, 1992: 116). The feeling of "we-ness", which had already been broadcast across countries and even continents by radio, was now visually beamed into the living room. According to Arnheim, this visual extension of our social network could bring about the awareness of "the place where we are located as one of many", making us "more modest, less egocentric" (Arnheim, 1957: 194). From these words a strong belief can be inferred, a belief that the introduction of electricity, the conquest of the ether and the use of the broadcasting properties of radio and television would eventually lead to a society free from prejudice, misunderstanding and all other unwanted characteristics (Spigel, 1992: 110–111). This belief corresponds with what has been described as the utopian idea of ideal communication.

But the reality of television history has proved otherwise. The television industry's decision makers were often men with a background in radio and they were determined not to make the same mistake again, which was to underestimate and overlook the economic value to be gained with an entertainment-driven broadcasting model (Boddy, 1990: 16). The television viewer does not have complete control over what images are projected on his screen, as foreseen by Albert Robida (1883) in his imaginative *Le Vingtième Siècle*, but instead has to wade through whatever is offered. Despite the consciousness-raising characteristics he attributed to television, Arnheim himself was not overly enthusiastic about the prospect of everyone having to use a television to connect to the outside world: "The pathetic hermit, squatting in his room, hundreds of miles away from the scene that he experiences as his present life, the 'viewer' who cannot even laugh or applaud without feeling ridiculous, is the final product of a century-long development, which has led from the campfire, the market place and the arena to the lonesome consumer of spectacles today" (Arnheim, 1957: 198). The explosion of channels has made it very unlikely that we all share the same television experience, thereby making a shared feeling of "we-ness" less likely to occur (with, maybe, the exception of single occurring or major sporting events, or breaking news). Although television might have corrected the balance of the extension of our sense organs, it has not brought us very much closer to angelic communication.

When looking at media evolution, we clearly see the process of remediation at work. The telegraph disconnected the transfer of information from its physical carrier; the telephone added natural speech to the telegraph; radio freed the telephone of its wires; and television added images to audible communication. Each addition can be seen as trying to realise ideal communication; that is, elimination of distance and time, making it possible to contact anyone at any time, in whatever way our sense organs would normally allow. Each attempt failed, however, not only because of imperfections in the medium itself, but also because social, cultural, political and economic factors determined otherwise. It is exactly those failures that motivate us to keep on searching. The next section looks at how the rise of mobile telephony continues this search.

1.4 Mobile Telephony

As discussed, it is an intrinsic human desire to bridge all chasms by realising angelic communication, making it possible to share one's thoughts or be present in each other's company at any time and any place. This is not to say that each individual harbours the wish to be always connected, but rather that everyone has, at one time or another, experienced the feeling that things could have been better with "ideal" communication. As a result, the process of remediation is continuously spawning new media with new functions, and it is in this process that we find the roots of mobile telephony. With the expansion of the fixed telephone network, it was soon discovered that not only bankers, lawyers or doctors appreciated the benefits of easily accessible communication, but also the average person. One important condition for being able to connect to one's friends and family was of course to have a telephone connection, but this was quickly met when the technology became cheap and the number of providers grew.

There were disadvantages to the system, of course. One of the more obvious was that in order to use the phone, one had to be at home, or at the office, or at any place where the wires of the fixed network ended in one's personal connection. There was no way of using a telephone while on the road and no way of making the telephone system itself find out which instrument was nearest to the person to be reached (Dick, a character in Woody Allen's "Play it again, Sam" (1972), circumvents this problem by using every fixed phone he happens to see to let his secretary know on which numbers he can be reached that day). For most casual users this was not a big problem, but it did not help to fulfil the promise of communication by anyone, anywhere. Another drawback of the growing network was the increasing amount of wiring needed above and below ground. Once again, the imperfections of the medium prompted the search for better solutions.

The next subsections will describe how mobile telephony has developed, what was expected with the idea of ideal communication in mind, how current use tries to realise those expectations and in what ways the dream has still not been fulfilled.

1.4.1 Development

The beginning of the 20th century marked an age in which the telephone was hailed as the successor to the telegraph. In the same period, the first successful experiments with radio telegraphy were performed. To many scientists it was therefore a logical step to try to combine these two technologies and realise wireless telephony. One of the first but lesser known figures involved in the earliest wireless telephony experiments was A. Frederick Collins. Claiming to be the "Inventor of the Wireless Telephone [in] 1899" (Collins, 1922), he wrote an impressive number of technical articles on wireless telephony during the first 10 years of the 20th century, heralding the end of telegraphy and wired telephony. He never used his company's money to mass-produce wireless telephones however, and was sentenced to three years in jail after being found guilty of giving a fraudulent demonstration. The first recognisable outcome of successful wireless experiments was, as we know, radio. Although its main function eventually turned out not to be establishing point-to-point contact (with some exceptions), radio's property of transmitting voice and sound signals without the aid of wires was clearly recognised.

Navies were the first to benefit from this liberation from wires. Before the arrival of radio communication, ships had to rely on flag or light signals when navigating or exchanging information. During heavy weather or mist, this posed almost insurmountable problems (Douglas, 1987: 265–266). Not only naval officers saw radio's potential for establishing wireless point-to-point contact, however. With the arrival of the fixed telephone network, police forces were expected to receive an increase in reports, so it became vital to reach police officers on the beat as quickly as possible. As early as in 1910 an idea was put forward to provide each policeman with his own personal telephone number, so he could be reached wherever he was (Pool *et al.*, 1977: 138). This idea took almost 10 years to come to fruition. Between 1921 and 1928, Robert L. Batts worked for the Detroit Michigan Police Department on a mobile radio installation, which could be built into a car (Waveguide, 1999; Slomnicki, 1999). The system was one-way only: policemen could be reached, but they had to get out of their car to find a telephone booth to call back.

Even with its initial drawbacks, the Detroit experiment with mobile radio communication was a great success and was followed and improved on by many other police stations. Within 10 years, the one-way system had been replaced by technology that used a so-called "push-talk" principle: a button had to be kept pushed in order to talk, and released to listen. In 1969, the Improved Mobile Telephone Service (IMTS) was introduced, which eliminated push-talk and finally made real mobile telephone conversation possible. Mobile telephone systems relied on transport by a vehicle, until the invention of the transistor, which led to the age of miniaturisation. In 1973, Martin Cooper, a scientist working at Motorola, presented the first mobile phone that could be carried by hand. Commercial mobile telephony was launched in the USA in 1983, after a new Advanced Mobile Phone Service (AMPS) standard was established. In Europe, standardisation took a little longer to develop, and it was not until the mid-1990s that the Global System for Mobile Communications was created by the Groupe Speciale Mobile (GSM). These so-called second generation (2G) phones are currently being replaced by telephones equipped with third generation (3G) technology, based on new high bandwidth mobile standards such as Universal Mobile Telephony Standard (UMTS), General Packet Radio Service (GPRS) and Enhanced Data rates for GSM Evolution (EDGE).

1.4.2 Expectations

Although Marconi's radio freed the telegraph from its wires, with subsequent work from such scientists as De Forest and Fessenden doing the same for telephony, radios were highly immobile at first and, on top of that, quickly lost their point-to-point behaviour to make way for broadcasting. Mobile telephony by radio thus did not materialise as soon as some would have liked. It was the desire to realise the dream of ideal communication that kept the momentum alive, resulting in today's mobile society. More than any other medium, the mobile provides its user with the ability to be a sender as well as a receiver of information, something radio and television do not offer to form a uniquely identifiable node in a local or global communication network, independent of his or her geographical position. This potentially constitutes a strong sense of connectedness with all other nodes: "Thanks to the mobile, you and the rest of the network will be wherever I am, and vice versa."

This vision of the mobile telephone is in line with what was attempted with previous media, namely to be "present" at places where one was not, to contact others when looking for help or a conversation and to extend the opportunity of experiencing Heidegger's notion of In-der-Welt-Sein (being-in-the-world) to the highest possible degree. Again, the underlying ideology is that with new media we will be able to reach and understand each other quicker and better than before, making the world a better place. This is clearly visible in advertisement campaigns for mobile telephony. Business meetings can be arranged or rearranged in an instant, a driver whose car has broken down can contact a car repair service and still make it to their destination, and when away you can always stay in touch with your loved ones. Almost without exception these advertisements stress the idea that, with your mobile, all problems can be solved and relationships will blossom as never before. In a more recent example, the Ericsson commercial "Into the Mobile Future" shows us, in succession, how people bridge space and time when they play chess, call emergency services, plan romantic dates, buy houses or snowmobiles and look up travel and tourist information; all by use of the mobile, and with a smile on their faces. Showing a happy user is of course one of the oldest tricks in the advertisement book, but by stressing that the mobile can solve any problem it becomes the Holy Grail of communication. It is what the industry recognises as a weak spot in our desire, and they play the game well. What the telegraph, telephone, radio and television could not completely realise will now be possible using the newest super medium.

1.4.3 Current Use

When looking at the development and use of previous media, we can distinguish two important changes in the 1920s concerning communication media paradigms. First, the specific broadcasting properties of radio, and later of television, meant that media were no longer considered as necessarily having to facilitate point-to-point contact. As a result, profit-driven broadcast entertainment became a major player in deciding the future of telecommunication (Flichy, 1995: 108–111). Second, while the telegraph and telephone were built and initially marketed by inventor-entrepreneurs working alone or with just a few assistants, this changed with the advent of big companies that could afford immense laboratories and large-scale promotional campaigns. Corporations, such as RCA, General Electric and Westinghouse fought for and won control over the airwaves and gained patents for the essential technology, prohibiting the average radio or television user from becoming a sender or broadcaster as well.

Both these paradigm shifts were followed by yet another change in the way media technology was perceived. The dawn of the electronic age with the invention of the transistor in 1947 would prove to have an enormous impact on the size, price and number of media devices. On top of that, the growth of computing electronics made digitisation of information possible, transforming every information-processing entity into a potential node of an all-encompassing network. These three new paradigms emphasise the informational and entertainment value of communication technology; powerful companies largely in control of development and marketing of new media; and the desire to compress and integrate media into the global information network can be said to exert great influence on media evolution today.

The mobile phone in 2003, therefore, was presented as a lot more than a tool for calling someone or for being called. By integrating a variety of functions in a single handset, the mobile telephone has become a multimedia information processor, seemingly fitted with an array of conveniences: "The mobile phone is becoming a personal trusted device, a life management tool for business, work and leisure. It will take on many roles: an anchor point, a digital navigator and a lifestyle accessory that will help individuals control and enrich their lives" (Nokia, 2002).

The word "phone" is increasingly replaced by "device", a personal digital assistant. In the mobile world, having control over your life means being able to cope with all possible flows of information, from node to node, network to network. "Communication" is replaced by "information management", be it for business, personal or entertainment purposes. In the mobile, we see the process of remediation in its most tight-knit form: sending and receiving text (remediation of the letter or fax), playing music (remediation of the radio or record/CD player), making small payments (remediation of money or credit cards), playing games (remediation of the game console), taking pictures (remediation of the camera) or accessing the Internet (remediation of the modem).

To communicate like angels means the ability to access the information network at any time, at any place. Today, this network extends widely over the Earth, although it does not yet cover it completely. Access via the use of mobile phones has become easier, though. To give some examples: in 2002, Japan had 55 million mobile phone users (out of a population of 127 million), China over 200 million (out of 1.3 billion) and the Netherlands about 12 million (out of 16 million). In Finland almost 90% of the population has a mobile, compared to 62% in the USA. In almost all of these cases, the number of mobile phones exceeds the number of fixed telephone connections, something that has also already happened in large parts of the African continent (BBC, 2001). On a global scale, mobiles are increasingly becoming a standard tool for communication (Greenspan, 2002).

Leaving all remediated extras aside, the mobile phone is of course still predominantly used for eliminating space and time in our search for presence and togetherness. The stories resulting from this use are often spectacular, such as those of people that were saved from death, but they can also be tragic, as we have seen with the 11 September terrorist attacks on the World Trade Center buildings in New York, during which people in the hijacked planes used their mobile phones to say goodbye to their loved ones, or tried to alert the authorities. Such stories show how the mobile is used to try and throw lifelines into the pond of networked contact. With the promise of 3G phones, capable of being "always-on" and of processing much more data, it looks like the dream of angelic communication is closer than ever before.

1.4.4 Critical Analysis

While slick advertising campaigns and provocative press releases may predict a glorious future within our reach, they cannot hide the numerous disadvantages and problems that arise with each new medium. Sooner or later, the desire for ideal communication meets the limitations of the medium, and it is the resulting realisation that our hopes have not quite been met that sets forth a new search. Most things promised never materialise. Instead of a realistic prediction of the future we are shown an "exercise in science fiction", as David Rodowick argues, which has to convince us that "capitalism, for centuries the source of so many of the world's social problems and inequities, can still be the solution, if we only let it again transform itself historically by unleashing the productive capacity of digital communications technologies" (Rodowick, 1998: 2). Technological progress, according to Rodowick, is meant to maintain the status quo of the balance between producer and consumer, not to realise Utopia. For in Utopia, there would be no need to buy new things.

One of the biggest problems lies in the rapid integration of mobile technology into our lives. Most of us are not adequately prepared for a society in which the sentence "We'll call" has replaced regular scheduling. These "approximeetings" form an increasing source of annoyance, especially when they end up in endlessly rearranged or even cancelled appointments (Plant, 2001). Social patterns have also been disrupted. Is it polite to answer the mobile phone in public? When should it be left off? Do we want to be part of half conversations that do not concern us? This "m-etiquette" has not yet found its final form and even when it has it is likely to differ between social groups. Instead of inheriting a better understanding of each other, we are faced with an even bigger chance of miscommunication and misunderstanding.

The desire for omnipresence of the mobile phone poses two major problems. One is that we are expected to have a mobile phone in the first place. Not only does this impose huge social pressure on the "mobileless" to act and get with it, it also makes us think everyone ought to be reachable all the time. The danger of being confronted with work while on vacation is yet another stress-inducing factor, in an age where the term "information overload" is not uncommon and the boundary between the private and public domains is quickly dissolving. (See Kopomaa (2000) for an account of the way mobile phones privatise the public sphere.) The second problem with omnipresence and the desire to contact anyone at anytime is that a Big Brother scenario is becoming more plausible day-by-day. Being able, like an angel, to know exactly what the other person means or thinks is the bright side of the story, but ethical and moral problems with that tricky concept called privacy lie on the other side.

Finally, one can even doubt whether mobile telephony is really suited for communication, in the original sense of the word (from the Latin communicare, to share). Establishing a "post-modern encounter" in his *Heidegger, Habermas and the Mobile Phone*, George Myerson (2001) juxtaposes the two German philosophers' ideas on communication with those he finds in the mobile discourse. Heidegger sees communication as a process of finding one's place among others, by determining what makes us different from each other. Habermas argued that communication supposes an interaction between two or more persons, in which ideas, thoughts and wishes are made known, and to which others can react. According to Myerson, this is not at all what the mobile telephone is used for. People in the mobile age communicate "to satisfy … wants. The mobile is the key to satisfying your wants generally" (Meyerson, 2001: 25). The medium has transformed into a device, a personal communication centre aimed at exercising control over what we want. It is not the desire to know one another, but the desire *per se* that the mobile fulfils (*ibid.*: 20–21, 26).

1.5 Conclusion

Looking back at media history, we see that media have evolved by coincidence, hard work, ingenuity, luck, persistence and vision, as well as by political decisions, economic measures, wars and social changes. These winding roads show how unpredictable media evolution can be. For example, it was once predicted that television would establish visual point-to-point contact, but now it mainly delivers passive entertainment.

Nevertheless, all media we have seen so far have one thing in common: they were initially perceived as trying to bridge space and time to such an extent that people would be able to communicate without obstacles and without misunderstanding. Fear of miscommunication and restoration of the natural balance of our extended sense organs is what drives us to improve existing media. Mobile telephony can be seen as the most recent attempt to reach the utopian ideal, which superficially seems closer than ever before. With mobile communication, it is possible to reach any other node in the information network, independent of one's geographical position. However, as Peters argues, the dream will never be fulfilled. It is exactly the awareness of these failures that keeps us trying again and again.

Taking this idealistic idea of angelic communication as the basic premise for the way media have developed, one could deduce that the specific succession of different media consists of necessary steps. Caution is advised here, for such a teleological vision would typify earlier forms of communication media as primitive, when compared to media that are chronologically as well as normatively closer to the ideal. However, as we have seen, this does not mean we can predict the future of media. We may be told that one day "we will ...", but history has so far proven otherwise.

References

Arnheim R (1957) A forecast of television. In: *Film as Art*. University of California Press, Berkeley.

BBC News (2001) African mobile phone use booms. http://news.bbc.co.uk/2/hi/business/1651950.stm

Boddy W (1990) *Fifties Television: The Industry and Its Critics*. University of Illinois Press, Urbana.

Bolter JD, Grusin R (1999) *Remediation: Understanding New Media*. MIT Press, Cambridge.

Briggs A (1977) The pleasure telephone: a chapter in the prehistory of the media. In: Pool I de Sola (ed.). *The Social Impact of the Telephone*. MIT Press, Cambridge (Massachusetts).

Castells M (1997) *The Information Age: Economy, Society and Culture, Volume 1. The Rise of the Network Society*. Blackwell Publishers, Malden.

Cherry C (1977) The telephone system: creator of mobility and social change. In: Pool I de Sola (ed.). *The Social Impact of the Telephone*. MIT Press, Cambridge (Massachusetts).

CNN Spotlight (2000) The birth of the hot line. http://www.cnn.com/SPECIALS/cold.war/episodes/10/spotlight/

Collins AF (1922) *The Radio Amateur's Hand Book: A Complete, Authentic and Informative Work on Wireless Telegraphy and Telephony*. Thomas Y. Crowell Company, New York.

Douglas S (1987) *Inventing American Broadcasting, 1899–1922*. Johns Hopkins University Press, Baltimore.

Elsner M, Müller T, Spangenberg PM (1984) The early history of German television: the slow development of a fast medium. In: Gumbrecht HU, Pfeiffer KL (eds). *Materialities of Communication*. Stanford University Press, Stanford.

Flichy P (1995) *Dynamics of Modern Communication*. Sage Publications, London.

Fischer C (1992) *America Calling: A Social History of the Telephone to 1940*. University of California Press, Berkeley.

Greenspan R (2002) Multiple, Global Increases in Mobile. http://cyberatlas.internet.com/markets/wireless/article/0,,10094_1480731,00.html

Hirsch E (1998) New technologies and domestic consumption. In: Geraghty C, Lusted D (eds). *The Television Studies Book*. Arnold, London.

Katz JE, Aakhus M (2002) *Perpetual Contact: Mobile Communication, Private Talk, Public Performance*. Cambridge University Press, Cambridge.

Kopomaa T (2000) *The City in Your Pocket: Birth of the Mobile Information Society*. Gaudeamus, Helsinki.

Levinson P (1997) *The Soft Edge: A Natural History and Future of the Information Revolution*. Routledge, New York.

Lewin K (1951) *Field Theory in Social Science*. Harper, New York.

McLuhan M (1964) *Understanding Media: The Extensions of Man. New American Library*. New York.

Myerson G (2001) *Heidegger, Habermas and the Mobile Phone*. Icon Books Ltd., Cambridge.

Moores S (1988) "The box on the dresser": memories of early radio and everyday life. In: *Media, Culture and Society*, Volume 10, No. 1. Academic Press, London.

Nokia (2002) Calling the next generation. http://www.nokia.ca/english/media/White_Papers/ White_Paper_3G.pdf

Peters JD (1999) *Speaking into the Air*. University of Chicago Press, Chicago.

Plant S (2001) *On the Mobile: The Effects of Mobile Telephones on Social and Individual Life*. http://www.motorola.com/mot/doc/0/234_MotDoc.pdf

Pool I de Sola (ed.). (1977) *The Social Impact of the Telephone*. MIT Press, Cambridge (Massachusetts).

Robida A (1883) *Le Vingtième Siècle*. G.Decaux, Paris.

Rodowick D (1999) An Uncertain Utopia – Digital Culture, 1998 (draft version). In: Pias C (ed.). *Dreizehn Vorträge zur Medienkultur*. Verlag und Datenbank für Geisteswissenschaften, Weimar.

Sconce J (2000) *Haunted Media: Electronic Presence from Telegraphy to Television*. Duke University Press, Durham London.

Slomnicki J (1999) *Communications: Where Did It Start?* http://www.911dispatch.com/information/ historycomm.html

Spigel L (1992) *Make Room for TV: Television and the Family Ideal in Postwar America*. University of Chicago Press, Chicago.

Uricchio W (2000) Technologies of time. In: Olsson J (ed.). *Allegories of Communication: Intermedial Concerns from Cinema to the Digital*. University of California Press, Berkeley.

Waveguide (1999) A Brief History of Cellular. http://www.wave-guide.org/archives/waveguide_3/ cellular-history.html

2

History Repeating?
A Comparison of the
Launch and Uses of Fixed
and Mobile Phones

Amparo Lasen

2.1 Introduction

The Ancient Greeks dreamt of a device that would enable people to talk over long distances without the need for an interlocutor. They called such a device a "telephone". The idea of direct communication over electric wire eventually arrived with Morse's telegraph in 1838. However, even when Bell finally invented the electric-speaking telephone in 1876, it still took some time to find a common use for the device. Even though its invention had been anticipated for a long time, it arrived without a clear and agreed purpose, and was received simply as a curiosity (Young, 1991: VIII, 1–2). The inventors had great difficulties finding a buyer for the patent (Aronson, 1977: 15).

Some aspects of the history of society's adoption of the fixed-line telephone and the corresponding adoption of the mobile telephone will be discussed in this chapter, highlighting the differences and similarities of these histories. Even after allowing for the differences of social contexts and technical devices, the knowledge of early practices, conflicts, fears and hopes about telephones will nevertheless improve the understanding of the uses and social roles of mobile telephones. The interest of the comparison is to give an insight into what happens when new services and new devices enter a marketplace. The evolution of fixed-line telephones, people's practices and meanings, and the attitudes, beliefs and behaviours of the industry can provide valuable information concerning the evolution of mobile telephones and the launch of new services. This contradicts the common view in the information and communication technologies industry, that the future is not built by looking at the past. But this chapter tries to show that, according to other well-known adage, if one ignores the past one is condemned to repeat it.

Though the field of social studies about mobile phones is growing, such studies are still rare compared to other technologies. In the early days of the landline telephone media showed the same lack of interest. The period of controversies, polemics and astonishment after the invention of the telephone in 1876 was shorter than for other devices, such as the radio. This silence and absence of public debate reflects the quick adoption of the telephone as a taken for granted element of everyday middle-class life in the countries where it was first introduced. When the sociological studies about technical inventions arose in the 1930s at the University of Chicago, under the influence of William F. Ogburn, the telephone was already an old communication technology, lacking the excitement of radio and cinema.

Social scientists were not interested in landline phones when they were introduced, and the same has happened with mobile phones. As with the old phones, the speed of acceptance and "naturalisation" of mobiles are the reasons given for this lack of interest. Radio and cinema yesterday, and the Internet today are more spectacular and exciting subjects for scholars. Mobile phones are also elusive to conceptualise. Although an example of new media, they have the transparency of the fixed phones, in that as conveyors of speech their mediation is forgotten. It is necessary to find concepts to apprehend mobile phones and able to understand a technology that connects local (conversations) and global (network, satellites and transmission points), which is a point of contact between different domains (public and private, work and home). A "theoretical mobility" is required to study this assemblage of people and technology forming a network, a shared agency that enables actions at distance (Cooper, 2001: 29).

Technological change results from the struggles and negotiations among interested parties: inventors, producers, different users and governments. The history of the social role of the telephone is less the technical evolution of the telephone system than a series of arenas for negotiating issues crucial to the conduct of social life: who is outside and inside, who may speak and who may not, who has the authority and may be believed (Marvin, 1988: 3). The adoption by society of new technological devices is a relationship of mutual shaping, where technology accommodates, but also transforms, existing social practices. The telephone use, like the uses of other technological objects, is a constructed complex of habits, beliefs and procedures embedded in elaborate cultural codes of communication. The uses are a distribution of competences and performances between people and devices, the result of how people project their respective social worlds onto technologies and what their justifications and fears are.

The topics examined are first the telephone as a broadcasting service, and then the issue of the early adopters and the appropriate uses according to the industry. The views of the industry about the right users and uses affected the development of the market. Also presented is the impact of the telephone on household and work management. In addition to its functionality, the telephone presents a "fun" side; it is an "electronic toy". The comparison also exposes the social skills created by the use of the telephone, the disturbing aspects of new technologies reflected in health and social fears associated with the use of the devices, and how telephones are related to the sustaining of community links and social networks.

2.2 Broadcast Services

Since the 1880s, the telephone has been a carrier of point-to-point messages between individuals, but then it was also a medium of multiple address for public occasions: concerts, theatre, sports, church services and political campaigns. This use as a means of entertainment and broadcasting of news was one of the main uses of the device until the end of the 19th century. The broadcasting of news was both professional and improvised. In the first case, the content was provided by the phone companies or other private broadcasters; in the second, the content was self-created by the users. The telephone companies broadcasted news, concerts, weather reports and even informed their subscribers of the entry of the USA into the war against Spain in Cuba in 1898. This was a supplementary service offered to their subscribers in order to make the device more attractive, which presents many similarities with the alerts received by mobile phones nowadays.

Occasionally private companies were created, such as the Telefon Hirmondó in Budapest, which existed from 1893 until World War I. The subscribers received a full schedule of political, economic and sport news, as well as plays, lectures and concerts (Briggs, 1977: 40–65; Marvin, 1988: 223–231). In the USA a similar example was the Telephone Herald of Newark (Marvin, 1988: 228–230). Like ancestors of the radio, they broadcast a whole range of news, lectures, theatre and music. The diffusion of news and gossip was also improvised through party lines. These were collective lines shared by a several homes. People who shared the same line exchanged and asked for news, or maybe more often just eavesdropped in order to be up to date with the current issues (Marvin, 1988: 222). The use of telephone as an entertainment form also involved teleconferences for clubs and associations. The telephone as a broadcasting medium coexisted with its use as a conversational instrument from the 1880s, but progressively disappeared before the invention of the radio and therefore did not compete with this new medium (Burrows, 1924). The audiences attracted by most of the commercial efforts to broadcast through the telephone were very small. They were the early adopters, mostly from the upper classes, and the content of the programmes transmitted reflected their tastes and interests. Nowadays, the multimedia possibilities of Wireless Application Protocol (WAP) and

Table 2.1 Private broadcast medium

1880–1920	2000
• Concerts	• MP3
• Theatre	• Radio
• News	• Alerts
• Sport	• Sport
• Church services	• Clips (video)
• Political speeches	• Chatlines
• Weather reports	
• Teleconferences	
• Improvised broadcasting through party lines	

third-generation (3G) mobile phones resume this use of the telephone as a broadcasting tool, but when a mass market has already been reached. A better understanding of the failure of broadcast services via fixed lines could help in making broadcast services in 3G succeed (Table 2.1).

2.3 Early Adopters and Appropriate Use

The telephone descended from a "parent" technology, the telegraph. Hence initially the people involved in supplying it and the marketing policies they used were the same. This inheritance shaped its early history. The first use as a conversational device was as a substitute for private telegraph lines, operating just between two fixed points, usually the home and the business place. Telephone went from the one-way process of the broadcasting use to a restricted one-to-one dialogue. It started as an expensive device favoured by the upper classes, as is often the case with new technologies, and then grew popular with farmers in the USA and urban middle classes. For the first decades of its existence, use of the telephone was a businessmen's monopoly. It substituted for the telegraph in the commercial and professional communities. From the beginning, the marketing campaigns were aimed at educating the public. They try to suggest purposes, instruct people on how to use the telephone, provide new etiquette rules and nurture goodwill to the industry. The industry men of the early days had a misperception of telephone users (Fischer, 1992: 60, 62, 78, 85). The late introduction of sociability as a marketing point does not mean that it did not exist before. It is, rather, that the experts, the industry men, considered these uses inappropriate. Fischer (1992: 81) explains this fact in relation to the inheritance from the telegraph. Those former telegraph men considered the use for social conversation an abuse or a trivialisation of the service. It was considered "chit-chat" and "idle gossip". Inside and outside the industry, many people considered idle conversations an inappropriate invasion of the household. There were also worries about inappropriate contacts between men and women of different classes and about the loss of privacy.

The dismissing of women as incompetent users was also extended to black people, immigrants and farmers. These considerations, expressed for example in technical journals and the electrical press, as Marvin (1988: 17–32) shows, are part of the invention of the expert and the stigmatisation of lay people. They did not derive from the use of the device, but from the existing relationships and perceptions of the different social groups. They distinguish the outsiders and the insiders of the technological world. In the case of mobile phones, the invisibility of teenagers and youngsters, and the consequent surprise when they became active users that helped to make Short Messaging Service (SMS) popular, is a similar phenomenon. Paradoxically, the focus on teenagers' uses afterwards, in social research and in the media too, does not take into account that adults also share some of these uses, and that other age groups, such as elderly people, are also adopting the device and changing their patterns of usage with the time. These examples of distorted perception of the uses of a technological device, by those who produce it, prove that the promoters of a technology do not necessarily know or decide its final uses. Consumers develop new uses and ultimately decide which will predominate. These vendors and marketers are constrained not only by technical and economic attributes, but also by an interpretation of its uses that is shaped by their histories. The industry men are as deeply involved in the realm of cultural production as in the technical one.

According to Marvin, the pioneer telephone men were "a self-conscious class of technical experts seeking public acknowledgement, legitimation and reward in the pursuit of their task". The effort to invent themselves as an elite, to command high social status and power was focused on technological literacy and on special symbolic skills as experts. They had to distinguish themselves from mechanics and tinkerers, and from the enthusiastic but electrically illiterate public, by elevating the theoretical over the practical, the textual over the manual and science over craft (Marvin, 1988: 61). One of the surprising consequences is that the industry men did not consider the telephone a product for mass distribution during at least the first 50 years of its existence.

Looking at the American case, as Fischer describes it, it is highly remarkable that income determined whether the urban American subscribed to telephone as strongly at the beginning as 40 years later. The stagnation of the diffusion of the telephone down the class system contrasts with that of the automobile. Not only were phones a less exciting purchase, they were also much less cost effective. This stagnation seems to be attributable to the scepticism of the industry about farmers, working class, ethnic minorities and migrants. The attitude of the industry was not only reflected in the marketing strategies, but also in the quality of the service provided and in the accessibility. Rural areas, the Southern states, were long forgotten in the building of new wire systems and exchanges. However, even if the telephone was targeted at the urban North originally, telephones diffused most rapidly in the Midwest and West. Farmers were more likely to subscribe than city-dwellers, at least in the first two decades of the century. The reason for the American farmers' interest in the new device was their isolation. However, many subsequently gave up their telephones. The explanation provided by Fischer is that other technologies, such as the automobile, better fulfilled their needs to break their isolation, due to the poor telephone service they were offered. Although the industry was not effective in creating needs and shaping use, it did set the structure within which consumers could exercise choice. As one of the Bell Canada managers quoted by Fischer said in 1902: "Of the 60,000 people in the city not more than 1,200 have or require telephones … Telephone service is not universal in its character and should not be supported by tax money". This claim reveals Bell leaders' sincere conviction throughout the era from the telephone's invention to 1940 that the telephone was not for the masses (Fischer, 1992: 107–108, 120).

The industry views were, actually, a major obstacle to its own commercial interests. Fischer argues that the conservative view of telephone use manifested by the industry in the early days reflected the inheritance of the telegraph. The arrival of new managers without those links facilitated the change in the marketing strategies. A learning period for both experts and the public was necessary in order to develop new uses.

The desire of those in the industry to distinguish themselves from the other groups and to maintain social distances and privileges was an obstacle to the commercial goal of extending the use of the telephone; the success of which precisely resides in the number of subscribers, and the possibility to communicate with others. Being an element of social distinction in the early days explains the opposition, not only by the experts but also by the first subscribers, against a mass diffusion of the phone and the expansion of public phones (Marvin, 1988: 102). New technological devices often accomplish this function of social distinction. The same happened with mobile phones. In the 1980s and early 1990s, European and American mobile phone companies followed the same strategy as the fixed telephone companies in their early days: high prices, exclusive use and avoidance of mass market. In the Scandinavian countries, however, the marketing strategies targeted both the business market and the mass market. The distinctive, yuppie image was a nuisance to avoid as much as possible, because it gave the product a bad press. This use was perceived as a way of showing off. Advertising campaigns tried to counter this perceived relationship between the device and this use in order that people, who did not want to be identified as yuppies, were not ashamed of using and purchasing a mobile phone (Roos, 1993: 10, 12).

The passage to a mass market for mobile phones was faster than for fixed lines, but the history of its development was not. According to Barry Brown, it is almost the history of a non-development (Brown, 2001: 7–10). The first commercial systems were up and running in the 1940s, but it took 30 years to acquire a mass market. It was delayed by decisions to favour other technologies. The first prototypes were fairly crude technologies. The system suffered from a chronic lack of capacity. The frequencies used by a call could not be reused. They were blocked by one call for as far as the radio transmissions were received. The lack of frequency, not the lack of interest by the public, prevented the early mobile phones system from becoming mass-market devices. In 1976 there were 44,000 people with mobile phones in the USA, and 20,000 individuals on a waiting list of 5–10 years time. Nevertheless, the capacity problem had already been solved in 1947, by splitting the coverage area into individual cells, and the technological challenges arising from this procedure were solved by the late 1960s. However, mobile telephony was delayed throughout the 1970s. Regulatory and business decisions made by governments and by the telephone companies explain this delay. Authorities hesitated to allocate spectrum for the telephone system over the frequency required for new TV channels. Companies interested in the development of mobile telephony were involved in legal disputes with American Telephone and Telegraph (AT&T). Therefore, the first mass market for a commercial cellular phones system in the USA only started in 1983, 37 years after the first carphone service. This development was not much quicker in other countries without those regulatory constraints, such as the Scandinavian countries, where mobile phones were launched in 1981. Scientists and engineers were also not really interested in this technology, in contrast with the interest in the development of videophones in the 1960s and 1970s, despite the adverse results of market trials. The scientific preferences for one type of technology and the disdain for others, the strange attraction despite commercial failures, is called by Brown a form of pathology in the mind of technological developers (Brown, 2001: 9). It will be interesting to see in the coming years if the comeback in new mobile phones of the old idea of a videophone becomes a successful application or receives the same cold reception from most people, who dislike the idea of having to worry about their physical appearance and their surroundings when they are making a phone call.

Early adopters have similar culture, interests and knowledge as the engineers and the industry men. That makes it easier to target them. There is a common belief about the diffusion of new uses and habits from the elite to the mass. But in many cases, the perception of a product as an elite one, a "yuppie thing", or a mere business or work-related device, is a handicap to its mass adoption. Apparently the industry has difficulty addressing a multiplicity of different uses and users. As an example, nowadays teenagers, family and friends have eclipsed businessmen and workers from the mobile phone advertising and communication. Less fashionable users, considered marginal, such as elderly people or working class, are also out of sight of the industry and social studies about mobile phones.

The marketing campaigns and the geographical offer of services were not the only effects of the industry men's conceptions about the fixed-line phone. The companies regulated its accessibility and its uses. They dictated who could use it and what issues it was appropriate to discuss. Even in the case of emergencies there was no general agreement in the early days about whether popular channels of communication, such as the telephone, could be relied upon for reporting incidents instead of the official proceedings of the police in those cases. Similar reticence has also been shown towards mobile phones. In an interview with a French mobile user in March 2004, part of a comparative cross-national study included in the Vodafone Surrey Scholar project, the informant told how once his motorcycle broke down in the middle of the motorway. When he phoned the road police to ask for help and to give them his location, he was asked to walk towards the nearest emergency phone, placed by the road milestones. The policeman preferred to be sure of his location by the information provided by the emergency phone, rather than trust the informant.

In the 19th century the Bell Telephone Company removed the phones of those who allowed non-subscribers or "deadheads" to use them. Marvin (1988: 104) quoted some cases of customers who sued the company for that. The ownership of the device and the last word on who decided who could call, were matters of conflict in the early days. 3G services and their "adult content" will introduce again this question of whether companies should regulate the accessibility of customers to services and cut off those inappropriate users. Etiquette on the phone was also a concern for the industry men. Phone conversations revealed, in the early days, a relaxation in the common courtesy of speech, which was a class-based reference. The recommended good manners were those of middle-class intimacy: quiet voices, clearly enunciated words and dignified presentation. The importance and the anxieties about how to speak properly on the phone and what community of speakers was addressed in the reach of its wires can be seen in the example of the Ohio telephone company, quoted by Marvin (1988: 89), which removed the telephones from subscribers who used improper or vulgar language. Mobile phone operators have also published etiquette guides of mobile phone uses, trying without much success to influence users' practices.

An important lesson from the history of the landline telephone is the power of users to impose their own purposes and competence, and how neglected and marginal users find successful uses, unknown or dismissed before by the experts. Women and sociability, teenagers and SMS are two different examples. In the case of fixed phones, the passage from the early adopters to a mass market was slower than for mobile telephones. This delay was not due to a lack of public interest. Rather, the industry did not consider the telephone as a mass product. Even if the passage to a mass market was faster for mobile phones, in both cases the industry was surprised by the market's uptake (Table 2.2).

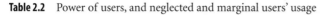

Table 2.2 Power of users, and neglected and marginal users' usage

1880–1920	2000
• Business and professionals were the first targeted market. Their use of the device for work matters was in accordance with the industry views.	• Business and professionals were the first targeted market. Their use of the device for work matters was in accordance with the industry views.
• Farmers' wives and other women used the phone mainly for social conversation and not just for household management. The commercial potential of this use was ignored by the industry for decades.	• Teenagers, but not only them, use SMS in a way that the industry had not foreseen.
• Targeting early adopters handicapped the spread of the telephone.	• Targeting early adopters, could it be a handicap to the development of new services?

2.4 Where Are You? Household and Work Organisation

As the industry and the early advertising campaigns highlighted it, the telephone introduced changes in the management of work activities. It helped the decentralisation of the office layout, and therefore contributed to the development of corporations and large organisations. Used according to practical requirements, it is a contribution to organised bureaucracy. It creates productive traffic (Cherry, 1977: 114), increases productivity by cutting the costs of acquiring information and co-ordinating schedules, and facilitates the control of the organisation resources, including the personnel ones. The telephone is a key element in the building of corporate empires. Apart from easing the violation of laws and the realisation of exchanges without leaving traces (Aronson, 1977: 32), phone communication facilitates the physical separation of the offices from the factories, allowing the managers to keep control of the production. Therefore, the telephone plays a role in the urban concentration of financial and business activities (Gottmann, 1977). It is also a central element in the work organisation and communication inside the skyscrapers, the symbols of corporate capitalism that arose at the beginning of the 20th century.

The effects of the mobile phone use in work activities are, unsurprisingly, mainly observed in the case of mobile workers. The monitoring activity carried out through the mobile phone is not only a way of controlling individuals on the move, it also allows workers to keep informed of what is happening in the office. By saying where they are and what they are doing individuals can simultaneously monitor their own and each other's work practices. Mobile phones are a means of remote background monitoring activity, which help with the catch-up period when returning to the office. Calls made to colleagues in the office are also a way of avoiding invisibility, of not being "out of sight, out of mind" (O'Hara *et al.*, 2001: 184). In our society the notion that individuals should be available and accountable to others, visibly and transparently at any time and place, becomes normal (Green, 2001: 33). The question "where are you?" (Laurier, 2001b) asked in mobile phone conversations is a form of establishing mutual contexts for communication, and enables shared circumstances between people communicating at a distance and a relation of mutual accountability and trust. Accountability is a feature common to co-present social relationships and to those established via mobile phones. Different technical systems (Internet and e-mail) allow people to communicate and to be monitored as well as information to be gathered by commercial institutions. Mobile phones are an example of this convergence of communication and information in the same device.

Laurier (2001a: 46–61) describes other ways of avoiding the undesirable consequences of the permanent accessibility provided by mobile phones in the case of mobile workers using call screening and voice mail. Mobile workers face a combination of two realities: the need to harmonise among multiple flows of activity and the interplay of planned and improvised action (Sherry, 2001: 112). They suffer the tension of the "anytime, anywhere" possibilities of communication of the ubiquitous mobile phone. The potential disruptions of the constant availability make it necessary to have a way of controlling the access. Mobile technologies of connectivity, like mobile phones, produce tensions in bringing together what is present and what is not. The voice mail service is a form of call storage, which is translated in return calls. In the case of the mobile worker studied by Laurier, the recorded calls are transformed into "Post-it" notes, emotional clues for the day ahead and requests to be noted and ignored. This service offers the possibility of responsiveness, without being drawn into the particular "concertedness" of phone conversations in "real time". It allows people's time space to be extended, orienting them towards distant and non-immediate requests and responses. Call screening is a finely crafted skill among mobile workers. It becomes even a mark of business credibility. "Everyone is busy call screening everyone else. If they do not, then Sylvia (the mobile worker studied by Laurier) wonders what is wrong with their business!"(Laurier, 2001a: 54). Churchill and Wakeford (2001: 174) understand mobility in the case of mobile workers as a continuum from tight mobility to loose mobility. This mobility is the experience of needing real-time synchrony while on the move, maintained throughout ongoing negotiations in established relationships where location information is easily shared or predicted. This is highly collaborative. Loose mobility concerns the requirement of accessing documents or information on the move, asynchronously and without requiring input at such detailed level. It is also highly co-operative, but not collaborative at a detailed level.

Telephones also play a role in the organisation of household activities and personal relationships (Dimmick *et al.*, 1994; Licoppe, 2002). The question

Table 2.3 Telephones' role in household activities and personal relationships

1880–1920	2000
Work	
• Changes in corporate space: monitoring factories from the office.	• Inversion of the relations between the caller and the receiver: reciprocal monitoring between office and mobile workers.
• A contribution to the organised bureaucracy: – Cut the costs, in time, money and effort, of acquiring information and co-ordinating schedules. – Control of the organisation resources, including the personnel ones.	• Mobile calls to the office killing the laptop and the Personal Digital Assistant (PDA), when workers call their colleagues and secretaries in the office asking for information and documents instead of carrying them.
Home	
• Calls between home and work: husband and wife.	• Calls between home and work.
• Management of the household: shopping and invitations.	• Parents and teenagers: monitoring and resistance.

of accessibility is therefore directly linked to the mobile phone as a tool of surveillance or monitoring. In the example of the relationship between teenagers and their parents, the "mobile parenting" (Oksman and Rautiainen, 2002; Kasesniemi, 2003), the monitoring and regulation by adults, is supported and resisted by teenagers themselves in moves toward independence and control of their own affairs. Mobile phones provide a site of negotiation for monitoring, regulation and mutual accountability. Teenagers are aware of the importance of mobile phones with regard to safety and emergency situations. But they also avoid parents' monitoring by not answering the phone or not telling the truth (Green, 2001: 38–39; Ling, 2001). Social normalisation of monitoring practices at the level of everyday life in public and domestic settings means that individuals can use their mobile phones to assist their own surveillance by institutions, as well as resisting it. At the same time, they also engage in routine monitoring of themselves and each other through that same technology, and assume that others are self-regulating and accountable for their use of the devices in both co-present and tele-present contexts (Green, 2001: 43–44) (Table 2.3).

2.5 Electronic Toy

In the beginning, the telephone was considered a kind of "electrical toy", presented by Alexander Graham Bell as a new marvel of science. His shows were intended to be demonstrations of utility, in order to convince the audience that the device worked and then try to persuade them to pay for it. Those demonstrations involved the broadcasting of music, theatre and information. In 1877 an event at the Sunday School of Old John St. M.E. Church included recitations, singing and an exhibition of "Dr. Bell's Speaking and Singing Telephone". When an audience of bishops and priests in Quebec City in 1877 heard a voice singing: "Thou are so near and yet so far," they stood up and sang back into the telephone. These stunts created considerable publicity as newspapers relayed them around the world.

Mobile phones also have the playful aspect of the early days of the fixed telephone. Their meaning is not only practical and useful, but also affective and entertaining. Affective aspects of mobile phones uses go beyond their playfulness and have been studied by several authors (Fracchiolla, 2001; Beckers *et al.*, 2002; Lobet-Maris and Henin, 2002; Rivière, 2002; Ellwood-Clayton, 2003; Harper, 2003; Jauréguiberry, 2003; Lasen, 2003b, 2004; Vincent, 2003; 2004). Mobile phones facilitate creative expression, especially in the case of SMS and Multimedia Messaging Service (MMS). They are a kind of toy and tool for play, with games, animations, pictures, smileys, ring tones and MP3. According to Kopomaa (2000: 70–71), the spread of technologies goes from novelty to invisibility, but in contradiction to this stage of "sobering up", the playful attitude towards mobile phones is likely to survive into the future. The advertising of picture phones and the marketing of 3G highlight the fun and entertaining aspects. Findings of a comparative study in London, Paris and Madrid in 2004 show that a category of pictures taken with camera phones are the result of playing with the phone in a way they would not do with a usual camera: taking snapshots without seeing what is being taken, like holding the phone out of the car window and shooting, taking pictures of strangers on the bus. This adds a thrill, an emotional intensity, to the boring routines of daily commuting. Other instances observed in that study that reveal a playful use of the device are taking pictures of friends just to have a laugh, sending jokes by SMS or teasing with SMS and MMS. Other entertaining practices are facilitated by other applications, such as being able to listen to the radio or downloaded tunes. In one instance observed in London on a rainy afternoon, a couple of teenagers were playing the same beat with their phones, whereas a third friend was rapping some lyrics.

Mobile phones inspire users to play with them, and that is precisely one of the qualities that attract people to the device. The symbolism of miniature objects emphasises this personal and playful character. Miniature objects are assimilated into toys and to the nostalgia of childhood, and also to intimacy, mobility, secrecy and control. Recent observation carried out in London, Paris and Madrid, revealed that a growing number of people, women and men of different ages, keep their phones in their hands when they are not using them, fiddling with the device, almost cuddling it.

The accent on the playfulness of new technologies helps to make them familiar. It helps the public to learn how to use them, but also has its

Table 2.4   Playfulness of new technologies

1880–1920	2000
• Fairs and shows where the telephone is presented as a form of entertainment. • Newspapers relayed what happened in the shows giving publicity to the new device.	• High street: retail as a place to play? • Media publicity about the fun aspects of new devices.

disadvantages. A device labelled as a toy misses other opportunities. For instance, the promotion of picture phones and 3G as toy telephones can discourage their application in the working environment. Our research in London, Paris and Madrid found that the media camera phones are associated with teenagers and youngsters, rather than with professional people (Table 2.4).

2.6 Disturbing Novelty: Health and Social Fears

Fears related to the use of the telephone concern health, social behaviour and social relationships. "Moral panic" often follows the introduction of new technologies. The idea of mobile phones making people more secure is counterbalanced by the health and social fears associated with telephones, which make their owners more vulnerable.

"As civilisation advances new kinds of diseases are produced by novel agencies which are brought to bear on man's body and mind" reported the British Medical Journal in 1889. The fear of health risks derived from telephone use also arose in the early days of its development. Even "strong-minded and able-bodied men" were considered to be susceptible because of the "almost constant strain of the auditory apparatus" in people who used the telephone frequently. Fears of changes introduced by this new technology in people's life were manifested in the belief that the telephone could be a source of bodily distress and unbalance. The body is the first frame to make sense of new experiences. The body is the centre of human experience and the most familiar of the communication modes. It constitutes the touchstone to gauge, explore and interpret the unfamiliar, the critical juncture between nature and culture. Mobile phones affect people's bodies and extend their world. Through our bodies we learn and become sensitive to what the world is made of, thus the body can be defined as learning to be affected (Latour, 2004). The more our body is affected, the more aspects it becomes aware of, and therefore the larger and more complex our world becomes. Mobile phones affect the way people talk and write, move and gesticulate, feel, walk and look or ignore their surroundings.

Early users of the phone were worried about "aural overpressure", nervous excitability, buzzing in the ear, giddiness, euphoria, neuralgic pains

Table 2.5 Telephones and health hazards

1880–1920	2000
• Aural overpressure	• Stress, work overpressure
• Nervous excitability	• Nervous excitability
• Insanity	• Insanity
• Addiction	• Addiction
• Contagion of infectious diseases	• Cancer
	• Alzheimer

and even insanity due to the constant ringing, as a result of an excessive use of the device (Marvin, 1988: 132). In those days people also feared the contagion of infectious diseases, either by the wires carrying virus, germs and bacteria or by using the same device as sick people. For example, in 1885 rumours spread in Montreal about an epidemic of smallpox being carried by people's breath through the phone (Young, 1991: 34). Some of the risks considered are the same in both landline and mobile phones; others are different, following the more feared diseases of each period. Also in the case of mobile phones, health fears not only concern the device and its users but also the phone masts and those who live and work near them (Table 2.5).

The association between sensational crime and new communication devices is not a new phenomenon. "It's a well-known fact that no other section of the population avail themselves more readily and speedily of the latest triumph of science than the criminal class" explained Inspector Bonfield to a Chicago Herald reporter in 1888. Mobile phone operators were reluctant to launch "Pay As You Go" cards as they feared such cards would be mainly used for criminal purposes. And it is true that mobile phones have became indispensable working tools for flexible mobile workers, such as drug dealers, who require to be contacted anywhere, anytime by their customers. Widespread mobile phone theft and the aggression that is often involved in the stealing of phones is another link between phones and crime, particularly youth crime, which make the owners of mobile phones more vulnerable. This also affects the way people use their phones. Many users avoid phoning in public places where they think there is a risk of theft (Lasen, 2003a). This fear is also related to the "value paradox" of mobile phones (Vincent, 2003). As people become more attached and dependent on their mobile phone and feel that they cannot live without it and all that it contains (phone numbers, messages and the potential of the relationships that mobile phones allow), its value is so great that they do not take it out or use it in certain places for fear of losing it.

In the early days of the landline telephone some communities banned the device because the effects of use were perceived as harmful for social relationships, a source of conflicts. Worries about eavesdropping were present since the beginning of the spread of the telephone, but so was the interest in doing it. Many of the first telephone lines used in America were collective, called party lines, when a few households shared the same telephone line. Each household had a particular ringing tone so that each would know when the call was for them. Thus, eavesdropping was as easy as it was tempting. The conflicts and quarrels arising from this practice were the reason why the Amish banned the device (Fischer, 1992: 241). The Amish and the Mennonites argued over whether the telephone was a theologically acceptable device or an intolerable worldly seduction. The Amish ended up banning the telephone due to the conflicts and disputes originated by eavesdropping in the party lines. They thought, "if that is the way they are going to be used we would better not have them".

Mobile phones are an "indiscrete technology" (Cooper, 2001). It is not that mobile phones facilitate forms of social indiscretion, but they do have the capacity to blur distinctions between ostensibly discrete domains and categories, such as public and private, remote and distant, work and leisure; just as fixed phone did with gender and class boundaries. These categories were already problematic before the appearance of mobile phones. The telephone offers the possibility of mixing heterogeneous social worlds. That is at the same time a useful opportunity and a dreadful threat of intrusion. The phone embedded the social risk of permitting outsiders to cross boundaries of race, gender and class without penalty. Nowadays, the worries concerning paedophiles and the growing concern about the protection of childhood also add the age boundary between adults and children. 3G will extend to mobile telephones the existent worries about the access to children to and through the Internet. Phones altered the customary orders of secrecy and publicity, as well as the customary proprieties of address and interaction. Well-insulated communities of pre-telephone days could not remain forever untouched by these developments, nor were telephone companies able to ensure that emerging telephone communities would remain within the limits of social decorum and work-related use (Marvin, 1988: 107). For instance, fixed-line and mobile phones facilitate courtship beyond parental control (Ellwood-Clayton, 2003) and promote infidelity but also helped to track down the adulterous. The most disturbing assault in social distance exploited telephone anonymity, abusive and obscene calls having existed since the early times (Marvin, 1988: 88; Katz, 1999: 231–278). Nowadays, SMS serves this purpose being a way of bullying and spreading rumours and reputations among teenagers. The spreading of rumours through SMS goes beyond teenagers' sociality. One of the consequences of the Chinese official "economy with the truth" about severe acute respiratory syndrome (SARS) was the spreading of rumour about hundreds of new cases through SMS.

It has been stated that, by increasing the amount of public communication, mobile phones reinforce the boundary between acquaintances and strangers, because they prevent people from talking to strangers in public spaces, reducing these small exchanges that support social communication. Such a statement has to be contrasted with empirical observation. Patterns of social interaction among strangers are different in different cultures. Londoners, for instance, did not wait for mobile phones to stop talking to strangers in trains or at bus stops. On the other hand, in southern countries, where one can talk to a stranger without being considered potentially mad or dangerous, and where the rules of civil inattention in public places never really ruled, this communication pattern seems to prevail despite the success of mobile phones. Civil inattention is a concept coined by Goffman, which refers to the ways in which people in public places show awareness of other people's presence without making them the object of particular attention. For example, by a mutual eye catching exchange with which one person admits seeing another, swiftly followed by the withdrawing of the attention "so as to express that he does not constitute a target of special curiosity or design" (Goffman, 1963: 84).

It is a common consideration that the use of mobile phones extends the intimate social sphere at the expense of the public, because it permits private communication in public spaces, which entails a privatisation of the urban space. This view seems to forget that private conversations have always taken place in public, and that public and private activities and exchanges can take place either in public or in private spaces. Friends and relatives chatting in a train do not only discuss weather or politics, and lovers kiss on park benches. Home can be the place for work or for having an anti-globalisation meeting. Subjective experience of the urban space does not mean its privatisation. The idea of Richard Sennet (1986) that dealing with private and intimate matters in a public milieu is a sign of an uncivilised society is not only very debatable, but also an outrageously Anglo-Saxon, self-centred statement. Mobile phones are indiscreet because their use blurs the boundaries between social spaces, like private and public. The coexistence of and potential friction between public and private are now material and observable phenomena. At the same time these uses reveal the limits of this separation. The use of mobile phones is sometimes considered an intrusion of the private world into the public sphere, and also as a resource for achieving privacy inside the home for the different members of the household, as in the case of the teenagers at home, or when one member of the household wants to have a secret conversation.

The separation between work and personal life is a 20th century concept. The one between public sphere and privacy is also a modern concept, especially characteristic of Anglo-Saxon and Protestant societies. According to Grant and Kiesler (2001), wireless technologies bring us back to earlier times when the boundary between work and personal life was less distinct. A specific feature of mobile phones is to facilitate more communication in transitional work settings, like hallways and lobbies, and in "dead" times. They are also employed to make communication in mixed-use settings, like cars or restaurants. The authors reveal the resulting paradox: in previous era increased mobility led to an increasing separation of work and personal place and life, but wireless technologies may be changing that.

The interruption of the immersion in private thoughts by mobile phone conversations and ring tones is viewed by some authors as the violation of a basic right associated with the use and enjoyment of the urban space (Kopomaa, 2000: 44). Urban public spaces have a soundscape made of different sounds and noises, among them mobile phone rings. In a space beyond our control, we certainly tend to create our own space. But personal space does not equal private space. The use and enjoyment of those spaces are not correlated with any supposed right not to be interrupted. The experience of strolling in urban spaces shows the opposite indeed. Different images and information constantly interrupt us. Moreover, the subjective daydreaming and reflections of strollers and commuters should not be systematically considered private thoughts. The French expression, *flâneurie*, this way of idle strolling and daydreaming, which involves the observation of people and social types and contexts, is a way of reading the city, its populations and its spatial configurations around the issues of the fragmentation of experience. The peculiar aspect of the *flâneur*'s way of strolling, as described by Baudelaire and Benjamin (Benjamin, 1973), is the gaze, the joy of looking. He observes, watches, perceives, classifies and daydreams. This is a way of resisting the indifference towards the distinction between things, typical of denizens of big cities. The interruptions and disruption of the street world are the fodder of this particular type of pedestrian.

The use of mobile phones is perhaps better understood by leaving the separation between public and private, and considering it as a way of introducing new practices and meanings in urban public places and inside the home. For instance, in the case of teenagers, the use of mobile phone at home, as in the bedroom at night for the ritual of goodnight messages, is not only a way of escaping the monitoring of parents or of achieving privacy at home (DWRC, 2000). It is, in their own words, a way of bringing college home, of communicating between the two main worlds of their everyday life. It is also a way of giving a new meaning to this nighttime when one is alone in the bedroom before falling asleep. The computer, the hi-fi and the personal stereo also contribute to transforming the home space, creating a personal, but not individual, territory (Table 2.6).

2.7 Social Skills

New social skills in professional and private life are created by the use of the telephone. As a consequence of the mutual-shaping dynamic between technology and society, the introduction of a technological device in everyday life activities requires an adaptation of social rules of interaction. For instance, the use of landline telephones simplified the formalities that ruled face-to-face conversation (e.g. opening sentences, polite forms of address). The telephone also facilitated new ways of organising time and space.

Table 2.6 Negative aspects of telephones

1880–1920	2000
• Decline of traditional forms of interaction like visiting	• Decline of traditional forms of interaction like face-to-face conversations
• Loss of interest in taking part in social activities	• Loss of interest in taking part in social activities
• Inconsiderate behaviour	• Inconsiderate behaviour
• Obscene calls or anonymity	• Obscene calls, abusive SMS or caller ID
• Crime: easier to commit fraud	• Crime: stealing and aggression
• Blur distinctions between:	• Blur distinctions between:
– Discrete groups: social classes and gender	– Discrete groups
– Domains and categories	– Domains and categories: work and home
– Public and private: home open to calls from outside, strangers, work matters, etc.	– Public and private: private conversations in public places

The notion of "anytime", associated with mobile phone use, refers to a linear conception of time, which is the translation to time of the geometrical conception of space. This is a time where all moments are equivalent and measurable, where one can be synchronous to any others regardless of the place, the hour and the activity. The use of mobile phones is an example of how the discipline based on the time organisation and on the strict planning of different activities is replaced by continuous accessibility (Ling, 2004: Chapter 4). The phone call is an advance arrangement, anticipates future meetings and prepares concrete proposals. The systematic use of time is replaced by systematic accessibility. Mobile phones permit an increased temporal efficiency by flexibility in the use of time. They allow the postponing and rearrangement of schedules, meetings and appointments. Nothing has to be agreed in precise terms any more. The conception and experience of a rhythmic time open to last minute changes and to continuous rearrangements between different activities, in work or leisure situations, is not originated by the mobile phone (Lasen, 2001). But the device facilitates their diffusion. The mobile phone makes the clock and the calendar unnecessary. It is a calendar and a clock. It becomes a navigation tool determining the co-ordinates of everyday living. These co-ordinates are colleagues, customers, friends, acquaintances and their schedules or the lack of them. The mobile phone replaces the clock and the calendar as the timekeeper of the linear time concept. However, it is also the tool for a rhythmic, more spontaneous and "last minute" time organisation.

Thanks to the fixed telephone, people increased their total conversations and the use of phones to arrange dates, trips and meetings suggests that calling assisted, even if it did not generate, many inter-person encounters (Fischer, 1992: 236–237). Maintaining personal relations by telephone became common in the middle class and farms of the 1910s and early 1920s. By this time Americans used the phone largely for sociability. This was truer for women than for men, for the younger more than for the older, for the gregarious more than for the shy. The communication modes displaced by the use of the telephone were telegrams and hand-delivered notes. It probably cut down the casual drop-in visits and helped to arrange other meetings (Fischer, 1992: 253).

Instead of competing with other technologies of communication, mobile phones allow different communication patterns (Harper, 2001): short messages and short calls, the possibility of transmitting a mood or a particular experience in real time or, in work situations, brief conversations primarily serving the function of making sure both parties agree with some brief discussion of status or progress. The facility to connect and the facility to cut the connection, to screen incoming calls and to shorten conversations, have made digital networks popular and highly used. In mobile phone use the threshold for making contacts become lower; even lower when texting. Mobile phones allow people to do things that are difficult to do when they are face to face, in situations emotionally charged as breaking up or asking out. In these cases technology facilitates the task, enabling people to elude some of the awkward social consequences these situations entail (Harper, 2003), although, in the case of dumping someone, at the cost of being considered coward, cheap and tactless. This way of avoiding possible face-to-face conflicts by using SMS not only applies to personal communication. In May 2003 the British injury claims firm Accident Group announced to more than 2,000 of its workers that they were going to be sacked by sending texts to their company mobile phones.

Being a mobile device was one of the main attractions when the product was launched and marketed to mobile workers. Different kinds of mobility are related to mobile phones; mobility of the user, the device itself and the services that can be accessed from different locations. But since the mid-1990s surveys done in Europe, such as the one quoted in Fortunati (2001), have shown that there was not a correlation between the mobility of the users and the use of the mobile phone. This is contrary to the opinion that mobile phones are instruments that enhance mobility. The force of attraction of the device was beyond the constraints of mobile work or those of residential mobility. When the device enters the domestic and leisure spheres it changes its identity and loses its connotation of being a mobile technology (Fortunati, 2001: 98). Some of those convivial uses reveal another kind of mobility, called micromobility by Weilenmann and Larsson (2001), the way in which an artefact can be mobilised and manipulated for various purposes around a relatively circumscribed, or "at hand", domain. Mobile phones are also a tool for local interaction, rather than merely a device for communication with others who are distant. This kind of use can be observed in the way SMS is employed by teenagers (Weilenmann and Larsson, 2001; Taylor and Harper, 2002).

47

Mobile phone communications make two different spaces parallel: the physical space where one is talking and the virtual space of the conversation. While using the mobile phone we are having simultaneously remote and co-present interaction. Using a mobile phone in a public place demands simultaneously handling the voice call and face-to-face interaction. This simultaneity affects the way we behave in front of strangers. The etiquette of phoning, that requires focusing attention on the conversation, can be in conflict with the etiquette of public behaviour. The body language of mobile phone users differs from the usual one of pedestrians. Their faces do not show reserve but reflect the content of the conversation, displaying emotions not usually shown on the street. They avoid eye contact with people around, often ignoring them. Therefore, they fail to produce the deferential glances exchanged between strangers. This conflict of etiquette influences the social acceptability of the mobile phone use in public. The simultaneity of the face-to-face interaction and the phone conversation also affects friends and acquaintances. Findings of a comparative study of mobile phone use in public place in London, Madrid and Paris show that users in different cities find different solutions for this situation (Lasen, 2003a). For instance, in Madrid they tend to include a third party in the phone conversation. They facilitate a sort of interaction between the phone conversation and the present situation, trying not to render absent those who face them. In Paris and London, however, when the users do not move away from those present, they interrupt the face-to-face interaction by avoiding eye contact, while those who are close pretend not to hear the phone conversation. The use of a mobile while with others can be a cause of annoyance, conflict and arguments if those present feel that the priority is given to the phone conversation. Therefore, restraining the use and shortening the conversations seem to be rules of etiquette in these cases. Being in company is a reason for switching the phone off for Londoners and Parisians. In Madrid the priority seems to be to always be available, therefore users keep their phones on in these occasions.

The use of mobile phones in urban places creates new urban practices and new meanings for urban space. The discomfort experienced by some of the users phoning outdoors is not as strong as to let a call go unanswered. Moreover, in the street users do not have to worry about other people overhearing their conversations. It was observed in Paris and London that some urban spaces constitute a kind of temporary phone zone (Lasen, 2003a). People stop there, make a call and resume walking afterwards. In front of big stores' doors, near underground entrances or on some street corners, one can see this kind of improvised open-air wireless phone booth. In these places several people are phoning, apparently unaware of others doing the same. This behaviour, which could be probably found in other cities, shows how phone users have created a new use for urban spaces. Streets are not only transitional spaces but also a place to stop and talk. The use of mobile phones gives new life and meanings to transitional spaces and times of our everyday life. A phone call turns a "no place", like a train station, bus, airport or road, into a third place for chatting and playfulness.

As a tool of arranging affairs and managing social relationships, mobile phones intensify the use of public space. These changes force reconsideration of the definition of correct behaviour in public places, and of what is annoying. The spread of mobile phone use diminishes the annoyance of the phone ringing and the phone conversation, as long as certain rules are respected, like answering quickly and not talking too loudly. The conflict of etiquette, the rules of interaction in public places on one side and the rules of mobile phone use on the other is an example of the mutual shaping between society and technology. The use of urban public places influences the use of mobile phones and is changed by them. That is the reason for the disapproval of the users who ignore the presence of others, are too noisy or talk about intimate matters. This is an example of how the rules of interaction in public places are applied to mobile phone use. However, the growing tolerance of this way of using phones and the widespread use of mobile phones in public places – where it was considered unsuitable or even banned, like restaurants, concert halls or cinemas – is proof that the use of a technological device can change the perception and use of public places and the way of behaving in such places.

Another example of social skills influenced by mobile phone use is the changing etiquette of phoning, particularly of phoning in public places. The unwritten rules of etiquette are revealed by phone users' non-verbal behaviour (body gestures, gaze and volume of voice) and by the reactions of others, not always silent, to mobile phone use in public places. The conventional patterns of non-vocal communication are "central to understand how unwritten rules of mobile phone use are constituted as a moment-by-moment emergent social phenomenon" (Murtagh, 2001). Mobile phone use also allows the redefinition of codes of human interaction, the renegotiation of norms governing social and emotional relationships, concerning courtesy, reciprocity, accessibility and the display of emotions in public contexts (De Gournay, 2002). Concerning this last aspect, mobile phones have become the medium for the "publicisation" of emotional fulfilment. They announce to our fellow travellers, pedestrian or work colleagues how much we are in demand, how full our life is (*ibid.*: 200). De Gournay interprets this possibility of renegotiating the norms as an elimination of the criteria for evaluating "social capital". Good manners, verbal skills, signs of refinement in dress and level of education, all these elements of social competence are not necessary when interacting from a distance with a mobile phone. By asserting that using a screen or a telephone nobody is required to display the slightest social competence, De Gournay fails to account for rules and etiquette that characterise mobile phone communication, and she reproduces a social fear already present in the early days of the landline telephone (Table 2.7).

Table 2.7 Telephones and social skills

1880–1920	2000
• Managing interruptions.	• Managing interruptions, attention and availability.

• Social etiquette:
  – Rules of how to speak properly on the telephone.
  – Appropriate times to call.
	• Social etiquette: rules of how to speak properly on the telephone in public places.
	• Avoidance of the awkward consequences of emotionally charged situations: breaking up, asking out and firing workers.

• Urban concentration.
• Expansion of a dimension of social life:
  – Frequent checking-in.
  – Rapid updates.
  – Easy scheduling of appointments.
  – Quick exchanges of casual confidences.
  – Long-distance calls.
	• Rural sprawl.
	• *Micromobility*: The way in which an artefact can be mobilised and manipulated for various purposes around a relatively circumscribed, or "at hand" domain, for example, inside the office.
	• Micromanagement of time:
	  – Phone call is an advance arrangement, anticipates future meetings and prepares concrete proposals.
	  – Increasing flexibility in the use of time: postponing and rearranging of schedules, meetings and appointments. The exactitude in the measurement of time is no longer necessary to co-ordinate activities. Punctuality ceases to be the virtue it used to be.
	• New ways of using urban and transitional spaces: no places become third places.

2.8 Community and Social Networks

One of the social fears associated with mobile phone use is its contribution to the dissolution of community bounds and exchanges. Telephones are a tool for collaborative interaction in the local environment, serving to strengthen and renew the membership in a community. They help to support and maintain the relationships of the people who deal with them. Marvin (1988) highlights that the most admired achievement of the telephone, the wonderful ability to extend messages effortlessly and instantaneously across time and space, reproduced without lost of content, was not linked to a genuine sense of cultural encounter and exchange at the time it was launched. Contrary to a utopian view of universal communication, those who controlled the new technology, like most of the Western white middle and upper classes at that time, dismissed different cultures as being deficient by civilised standards and even unable to communicate meaningfully.

The use of the telephone for sociability purposes did not fit the view of the experts because they dismissed those who used it mostly for that purpose, namely women. However, regardless of these views, conversation and sociability were the main uses of the telephone in the early days. Research quoted by Fischer (1992: 230), carried out in Seattle in 1909, reveals that half of the calls had some social content and 30% were idle gossip, at a time when only about one-third of Seattle households had telephones. Rural people, especially farmers' wives, depended heavily on the telephone for sociability until they had cars. Women called often for social purposes and frequently even for simply "visiting", as far as back as the 1910s. Conversation is an important social process, serving to maintain networks and build communities. This aspect was ignored by industry leaders, journalists and other male critics. Time-budgets filled out by suburban New Yorkers before World War II revealed that women spent four times as much time on the telephone as men did. In other parts of the USA, the higher the proportion of adult women in the household, the higher the chances were that it had a telephone. The industry view in the 1920s was that women, acting as "chief executive officers" of the household, should telephone to order goods and services. Women's social calls were considered a problem and they initially tried to suppress them. But women cultivated their own purposes. Farmwomen used the phone to sustain social activities and help create community bonds in rural areas; in urban areas, middle and upper class women used the phone for organisational activities, young urban women used it for courting.

In the late 1920s and 1930s a change occurred and telephone advertising increasingly depicted women using the telephone for sustaining social contacts and conversation. Fischer proposes a few explanations for this gender difference. The isolation of women from adult contact during the day, the fact that married women's duties include the role of social manager (appointments, events, keeping informed about relatives and friends) and that North American women are more comfortable on the telephone because they are generally more sociable than men. So the telephone fits the typical female style of personal interaction more closely than it does for men. From the first decades of the 20th century women used the telephone to pursue what they wanted, conversation, and were responsible for the development of a culture of the telephone as they instigated its use for purposes of sociability.

Mobile phone advertising campaigns also moved from the work and business orientation of the early days to friends and family communication, where women and youngsters are equally as present as men, until the point that work and business users have disappeared from the latest mobile phone ads for picture phones. The large uptake of mobile phones in Europe and East Asia shows the growing importance of social connectiveness in purchasing and using the device. The relevance for personal life and interpersonal communication rather than the social or professional utility of the device is one of the findings of our cross-national research on mobile users quoted above that is also revealed in other studies (De Gournay, 2002).

Former advertising campaigns focusing on work issues not only failed to account for other kind of uses but also provided an individualistic view of mobile phone use, which did not correspond to the real experience of mobile workers. Advertising campaigns addressing mobile workers emphasised the individual use of mobile phones in both senses: as the enhancement of personal skills and control over the environment. Workers are represented as individualist in their use of the technology. They are independent and work alone. The technological devices are not part of a collaborative network of devices and collaborations (Churchill and Wakeford, 2001). Nevertheless, the use of mobile phones depends on stable social infrastructures. Mobile workers often find themselves in a situation of waiting. This waiting – for events to happen, for flights and trains, for information and business deals or for other people to do things – is a common experience which demonstrates an interconnected network of working relationships, collaborators and the interconnectedness of devices (Churchill and Wakeford, 2001). Mobile phones help to strengthen social bounds not only in the realm of private lives but also in work environments.

The use of mobile phones is closely related to personal lifestyles. They have a value and act as a symbolic marker. A range of meanings linked to identity, sexuality and desire are attached to mobile phones. Cooper *et al.* (2000) analyse the role of mobile phones in relation to gay lifestyles. The young men interviewed consider the mobile phone as a means of maintaining real-time contact with significant others, a decisive element of the gay lifestyle. The interest of such statements does not lie in whether they are true or not. On one hand, there is not just one gay lifestyle; on the other, the convivial use of mobile phones to strength friendship bonds and to maintain personal networks is not exclusive to gay men. But the consideration of a common use as a particular and distinctive one, linked to a specific group and lifestyle, demonstrates the importance of mobile phones as personal devices becoming part of one's own personality and of the feeling of belonging to a particular community. A similar example can be found in the importance of mobile phones for clubbers who use the device to prepare the night out, to make arrangements to meet other people and to communicate with those absent from the club (Miles and Moore, 2004).

Teenagers' use of mobile phones reveals that they are more than an individual device (Weilenmann and Larsson, 2001; Green, 2002; Johnsen, 2002; Kasesniemi and Rautianen, 2002; Ling and Yttri, 2002; Oksman and Rautiainen, 2002; 2003; Skog, 2002; Taylor and Harper, 2002; 2003; Kasesniemi, 2003; Lobet-Maris, 2003; Weilenmann, 2003; Ling, 2004; 2003). They are tools for collaborative interaction in the local environment. The sharing of mobile phones helps to invigorate and sustain social networks. Weilenmann and Larsson, in their research with Swedish teenagers, identify different forms of sharing, from the "minimal forms of sharing" like showing, writing or deciphering SMS, to the "hands-on ways of sharing", as when several people take part in a conversation sharing the phone in a kind of multi-party talk. The borrowing and lending of telephones is a mark of trust and friendship also making the mobile phone a collaborative resource. These Swedish teenagers also share mobile phones with unknown people in order to make contact, as when storing your phone number in other's mobile phones. The comparative study of mobile phone use in public places quoted above (Lasen, 2003a), found that adult users in Madrid also have a collaborative use of the mobile, for instance sharing the phone conversation with those who are face-to-face. Camera phones also facilitate the collective use of the device. Taking and showing pictures increases the ability of the device to be shared, modifying the strictly individual belonging of traditional mobile phones.

All these empirical examples show that the personal character of mobile phones does not imply that their use is individual, and even less individualistic. However, the individualistic argument proves difficult to avoid in the minds of social scientists. It seems that it occurs with the link between mobile phones and individualism in the same way as with fixed phones and the process of modernisation. If one believes that we live in individualistic societies, the success of mobile phones can only mean that they are "perfectly suited to the ideology of an individualistic society", as Kopomaa asserts, forgetting all the examples he provides in the same book invalidating such a statement. Mobile phones are a tool for networking, for nurturing and maintaining social cohesion of groups, for creating "nomadic communities". 3G telephones open this issue of communities and social networks to the virtual communities. Until now those were based on e-mail exchanges and the use of Internet. There is a different perception of telephones and e-mail addresses. 3G will pose the question of the access by strangers to our personal mobile phones.

The study of Fischer (1992: 194, 220–221) reveals few changes in localism. People were not uprooted by the new technologies, locating their activities and interests somewhat more often outside the towns, but mostly they expanded their local activities. The telephone cannot be substantially credited or blamed for undermining localism in the early 20th century. As Willey and Rice argued in a monograph published in 1933, people used the telephone to increase local ties much more than extra local ones. Phone calling strengthened localities against homogenising cultural forces, such as movies and radio. It also enabled Americans to participate in activities more frequently and more easily outside their localities. As Fischer affirms, people called relatives long distance, made more trips to tourist spots, and followed their sport teams to more games. Although the balance of change was in the direction of the wider world it was not a weighty shift, not as substantial as the increase in total social activity. Mobile phones serve to increase the communication within a small group of people locally, mostly friends and family, instead of helping to enlarge the circle of friends and acquaintances.

First users perceived the fixed-line telephone as a way of reducing loneliness and anxiety, a purveyor of increased feeling of psychological and even physical security. Keeping in touch with the loved and the familiar allows one to conjugate security and geographical mobility (Cherry, 1977). The same applies to mobile phones. These characteristics were studied in a particular situation, the sudden deprivation of the telephone. In February 1975, a fire in a switching centre in New York left a 300-block area of Manhattan (around 100,000 customers) without phone service for 33 days. The research carried out afterwards about this period studied the hypothesis of the telephone as a reducer of loneliness and anxiety, a maintainer of the groups' cohesion, or as an intruder in the private sphere. The results showed that there is no satisfactory alternative to the telephone and its essential role in urban lifestyles. These unfortunate New Yorkers felt isolated, uneasy and less in control of their lives. There were not clear compensatory behaviours. The increase in TV and radio consuming or in visiting friends was modest. The incoming calls were missed more than the outgoing. The telephone is an instrument of urban adaptation. It allows an imminent connectedness and an immediate interaction that shapes the symbolic proximity, which counteracts social mobility (Wurtzel and Turner, 1977: 246–261). The possibility of perpetual contact allowed by mobile phones is so important that the loss of this capability produces strong feelings, such as panic when users lose or forget their phone, or anger against the operator when the network fails. This aspect outlines the importance of the resilience of networks for mobile telephone users. On 20th February 2003 the 8.7 million Vodafone subscribers in Spain could not use their phones because of a breakdown. The journalist who reported the problem in El Pais newspaper described how the subscribers were left "absolutely cut off for hours, unable to make or receive calls on their mobile phones". The expression "absolutely cut off" can seem exaggerated, as other ways of communication were available. But that was the main feeling of the users, not only those who depend on their phones for work and business but also for those who use them to be in contact with friends and family. These users interviewed on the TV news also felt isolated, uneasy and less in control of their lives, as in the New York case described above. Again, people's main worries were not about outgoing calls, after all there were other ways of making calls, but the loss of the ability to receive calls from the people they know. The company received more than 400,000 complaints and the regional governments of Catalonia and Andalucia threatened to sue the company.

As a person-to-person communication tool, phones, fixed or mobile, share the emotional aspects characteristic of human communication, the "emotional attunement" necessary for the achievement of human communication. Conversation, though banal in appearance, is an accomplishment achieved only through the co-operative and sensitive performance of the participants. Participants in a conversation show emotional identification with each other. There is a tacit agreement to help each other along, to accept each other's presentation, because the failure of one produces embarrassment in others. Thus, every conversational move is fraught with emotional potential. Anger, pride, anguish, confidence, embarrassment, guilt and gratitude are potential emotional accompaniments of conversation, no matter what the content is. The "forms of talk" in Goffman's words (1981), the conversation as a social form, makes the emotion.

As we know from the pioneer studies of Durkheim (1915/1965), the sharing of emotions is crucial to the creation and maintaining of social bonds. Bauman's description of the situation of the unbound contemporary individuals, forced to tie together whatever bonds they want to use as a link to engage with the rest of the human world "by their own efforts with the help of their own skills and dedication" (Bauman, 2003) gives a strong insight into the role of mobile phones in our societies, as a suitable tool to accomplish this task of providing connections. Mobile phones also respond to the other need outlined by Bauman, that these bonds should be "only loosely tied, so that they can be untied again, with little delay, when the settings change". Mobile phones and their promise of perpetual contact and permanent accessibility provide the assurance of connections, needing to be ceaselessly renewed. But they are also the witness and accountant of this anguishing situation where contacts and relations cannot be taken for granted: "Why didn't I get any messages today? Why didn't she call me back?" (Jauréguiberry, 2003).

Connections are "virtual relations", easy to enter and to exit. But networking becomes an all-consuming endless task, where mobile phones provide help at the same time that they multiply the occasions for being in touch, the calls and answers to manage. Paradoxically they can reduce the complexity of managing one's relational capital and networks while keeping the level of complexity by amplifying the number of contacts and opportunities to get in touch that one has to fulfil (Table 2.8).

Table 2.8 Relationships and telephones

1880–1920	2000
• *Women*: In the early days, women were the main group to use the phone to sustain community links.	• *Teenagers*: Their use of mobile phone is mainly related to friendship and sociability.
• Expansion of the local and extra local activities.	• Increasing of the number of contacts with a small number of people.
• Increasing of local ties.	
• Keep in contact with friends and family.	• Keep in contact with friends and family.
• Reducing of loneliness and anxiety.	• Reducing of loneliness and anxiety, but paradoxically phones become witness and providers of these same fears and anxieties.

2.9 Conclusion

The history of the introduction of landline telephones reveals that the affordability and the availability of the phone came first to the more affluent households, and much later to the working class ones. The prejudices, narrow-mindedness and will to keep the phone as an elitist product of the early industry men resulted in the slow spread of the telephone. It represented the expansion of a dimension of social life: the realm of frequent checking-in, rapid updates, easy scheduling of appointments and quick exchanges of casual confidences, as well as the sphere of long-distance conversations. The case of the early years of telephone use in the USA did not reveal any clear sequence of dramatic social changes. There were no psychological changes that mirrored the characteristics of telephone. Americans used the telephone to enhance the way of life to which they were already committed. Phones have a particular history, different from other technologies, such as the automobile. This history is also a proof of the users' relative autonomy from the pressure of vendors and from any supposed technological imperative. Such autonomy is never total. It is limited by structures beyond their control, such as income and costs, where companies marketed their services and the role of government (Fischer, 1992: 268, 269).

Mobile phones are the most successful communication devices ever. The commercial launch of the device took a long time from the creation of the first prototypes. The pattern of diffusion followed that of the landline phones, from business and upper classes to a mass market centred on sociability uses. But this was achieved in a much shorter period of time. The reasons for purchasing the device were also the same: work efficiency, security in emergency and sociability. Many of the studies quoted in this chapter concern mobile workers or teenagers, the former in order to analyse the use of mobile phones in a work environment, the latter to consider the creative uses and the importance given to mobile phones in their everyday life. Teenagers, youngsters and young adults have adopted the new device faster and more openly. According to the Eurobarometer survey of 10,000 Europeans between the ages of 15 and 24 carried out in April and May 2001, mobile phones top the list of new technologies used by young people at least once a week (80%), with very little variation between countries. This is far ahead of computers, the Internet and e-mail.

The launch and spread of wireless phones have aroused some fears, as happened with the landline telephones. A certain amount of "moral panic" about the effects follows the adoption of new technologies. Some of those fears are similar in both cases: threats to the health, danger of addiction, the decline of traditional interactions, the loss of interest in taking part in social activities or inconsiderate behaviour. Others are new, such as the privatisation of public space, the intrusion of work into the private sphere or the increased possibilities for control. It has been argued above that the relationship of mutual shaping between technology, phones in this case, and social practices produces a more complex outcome than a unilateral growing control and the privatisation of the public space. Mobile phones allow new practices and uses of urban and public places and contribute to exposing the erosion of the division between private and public already existent in contemporary society.

Some of the issues described could be useful for further developments in the field of mobile communication devices. One is the conflict between the users' views and the industry men regarding the ownership of the device and the right uses. This conflict between users and producers could be important in the case of 3G mobile phones and the increasing number of partners. Another important teaching from the history of the landline telephone is the power of people to impose their purposes, revealing the illusion of the industry aim to educate the users, even in a constraining context, and how neglected and marginal users find successful uses, unknown or dismissed before by the experts.

The comparison also reveals the tendency in social sciences to consider the uses of new technologies of communication as ways of deepening social trends believed to be dominant (individualism, reduction of the public sphere, weakening of communities and social networks). Empirical research about fixed and mobile telephony shows a very different picture. Phone communication contributes to the maintenance of local communities and close groups, and in some cases takes part in the organisation of social movements (Rheingold, 2002), such as those in Philippines and South Korea, public and politic actions, such as the anti-globalisation protests in Seattle, anti-war demonstrations in London, or demonstrations in Madrid and other Spanish cities before the 2004 general election. But phone communication is not utopian. It happens here and now and has to deal with the paradoxes and limitations of human exchanges and social bonds present in our society.

References

Aronson S (1971) The sociology of the telephone. *International Journal of Comparative Sociology*, **12**: 153–167.
Aronson S (1977) Bell's electrical toy: what's the use? The sociology of early telephone usage. In: De Sola Pool I (ed.). *The Social Impact of the Telephone*. MIT Press, Cambridge, Massachusetts; pp. 15–36.
Bauman Z (2003) *Liquid Love*, Blackwell, Oxford.
Benjamin W (1973) *Charles Baudelaire. A Lyric Poet in the Era of High Capitalism*. New Left Books, London.
Briggs A (1977) The pleasure telephone: a chapter in the prehistory of the media. In: De Sola Pool I (ed.). *The Social Impact of the Telephone*. MIT Press, Cambridge, Massachusetts; pp. 40–65.
Brown B (2001) Studying the use of mobile technology. In: Brown B, Green N, Harper R (eds). *Wireless World. Social and Interactional Aspects of the Mobile Age*. Springer-Verlag, London; pp. 3–15.
Brown B, Green N, Harper R (eds) (2001) *Wireless World. Social and Interactional Aspects of the Mobile Age*. Springer-Verlag, London.
Burrows AR (1924) *The Story of Broadcasting*. Casell and Co. Ltd., London.

Cherry C (1977) The telephone system: creator of mobility and social change. In: De Sola Pool I (ed.). *The Social Impact of the Telephone*. MIT Press, Cambridge, Massachusetts; pp. 112–126.

Churchill EF, Wakeford N (2001) Framing mobile collaboration and mobile technologies. In: Brown B, Green N, Harper R (eds). *Wireless World. Social and Interactional Aspects of the Mobile Age*. Springer-Verlag, London; pp. 154–179.

Cooper G (2001) The mutable mobile: social theory in the wireless world. In: Brown B, Green N, Harper R (eds). *Wireless World. Social and Interactional Aspects of the Mobile Age*. Springer-Verlag, London; pp. 19–31.

Cooper G, Green N, Moore K (2000) Mobile culture: the symbolic meanings of a technical artefact. *Culture, Psychology and New Technologies Symposium*, December.

De Gournay C (2002) Pretense of intimacy in France. In: Katz JE, Aakhus M (eds). *Perpetual Contact: Mobile Communication, Private Talk, Public Performance*. Cambridge University Press; pp. 193–205.

De Sola Pool I (ed.) (1977) *The Social Impact of the Telephone*. MIT Press, Cambridge, Massachusetts.

Dimmick J, Sikand J, Pattersons S (1994) The gratifications of the household telephone: sociability, instrumentality and reassurance. *Communication Research*, **21**: 643.

Durkheim E (1965) *The Elementary Forms of the Religious Life*. Free Press, New York.

DWRC (2000) The use of mobile phones by school children. *STEMPEC Project Report*. Unpublished.

Ellwood-Clayton B (2003) Virtual strangers. Young love and texting in the Filipino archipelago of cyberspace. In: Nyíri K (ed.). *Mobile Democracy. Essays on Society, Self and Politics*. Passagen Verlag, Vienna; pp. 225–235.

Fischer CS (1992) *America Calling. A Social History of the Telephone to 1940*. University of California Press.

Fortunati L (1997) The ambiguous image of the mobile phone. In: Haddon L (ed.). *Communications on the Move: The Experience of Mobile Telephony in the 1990s, COST 248*. European Commission, Telia AB, Sweden.

Fortunati L (2001) The mobile phone: an identity on the move. *Personal and Ubiquitous Computing*, **5**(2): 85–98.

Fracchiolla B (2001) Le téléphone portable, pour une nouvelle écologie de la vie urbaine? Esprit critique. *Revue Electronique de Sociologie*, **3**(6), www.espritcritique.org/0306/article2.html

Goffman E (1963) *Behaviour in Public Places. Notes on the Social Organization of Gatherings*. Free Press.

Goffman E (1981) *Forms of Talk*. University of Pennsylvania Press, Philadelphia.

Gottmann J (1977) Megapolis and antipolis: the telephone and the structure of the city. In: De Sola Pool I (ed.). *The Social Impact of the Telephone*. MIT Press, Cambridge, Massachusetts; pp. 303–317.

Grant D, Kiesler S (2001) Blurring the boundaries: cell phones, mobility and the line between work and personal life. In: Brown B, Green N, Harper R (eds). *Wireless World. Social and Interactional aspects of the Mobile Age*. Springer-Verlag, London; pp. 121–132.

Green N (2001) Who's watching whom? Monitoring and accountability in mobile relations. In: Brown B, Green N, Harper R (eds). *Wireless World. Social and Interactional Aspects of the Mobile Age*. Springer-Verlag, London; pp. 32–45.

Green N (2002) Outwardly mobile: young people and mobile technologies. In: Katz J (ed.). *Machines that Become us: The Social Context of Personal Communication Technology*. Transaction Publishers, New Jersey; pp. 201–217.

Haddon L (ed.) (1997) *Communications on the Move: The Experience of Mobile Telephony in the 1990s, COST 248*. European Commission, Telia AB, Sweden.

Harper R (2001) The mobile interface: old technologies and new arguments. In: Brown B, Green N, Harper R (eds). *Wireless World. Social and Interactional Aspects of the Mobile Age*. Springer-Verlag, London; pp. 207–226.

Harper R (2003) Are mobiles good or bad for society? In: Nyíri K (ed.). *Mobile Democracy. Essays on Society, Self and Politics*. Passagen Verlag, Vienna; pp. 71–94.

Jauréguiberry F (2003) *Les Branchés du Portable*. PUF, Paris.

Johnsen T (2002) The social context of the mobile phone use of Norwegian teens. In: Katz J (ed.). *Machines that Become us: The Social Context of Personal Communication Technology*. Transaction Publishers, New Jersey; pp. 161–169.

Kasesniemi EL (2003) *Mobile Messages. Young People and a New Communication Culture*. Tampere University Press.

Kasesniemi E, Rautianen P (2002) Mobile culture of children and teenagers in Finland. In: Katz JE, Aakhus M (eds). *Perpetual Contact: Mobile Communication, Private Talk, Public Performance*. Cambridge University Press.

Katz J (ed.) *Machines that Become us: The Social Context of Personal Communication Technology*. Transaction Publishers, New Jersey.

Katz JE (1999) *Connections. Social and Cultural Studies of the Telephone in American Life*. Transaction Publishers, New Brauswick, New Jersey.

Katz JE, Aakhus M (eds) (2002) *Perpetual Contact: Mobile Communication, Private Talk, Public Performance*. Cambridge University Press.

Kemper TD (1991) An introduction to the sociology of emotions. *International Review of Studies on Emotion*, **1**: 301–349.

Kopomaa T (2000) *The City in Your Pocket. Birth of the Mobile Information Society*. Gaudeamus, Helsinki.

Lasen A (2001) *Le Temps des Jeunes: Rythmes, Durée et Virtualités*. L'Harmattan, Paris.

Lasen A (2003a) *A Comparative Study of Mobile Phone Uses in Public Places in London, Madrid and Paris*. DWRC, University of Surrey, Guilford.

Lasen A (2003b) *Emotions and Digital Devices. Affective Computing and Mobile Phones*. DWRC, University of Surrey, Guilford.

Lasen A (2004) Affective mobile phones. *An Insight into How Mobile Phones Mediate Emotions Based on Fieldwork Carried out in London, Madrid and Paris. Proceedings of the Fifth Wireless World Conference*. DWRC, University of Surrey, Guilford.

Latour B (2004) How to talk about the body? The normative dimension of science studies. *Body and Society*, **10**(2/3): 205–229.

Laurier E (2001a) The region as a socio-technical accomplishment of mobile workers. In: Brown B, Green N, Harper R (eds). *Wireless World. Social and Interactional Aspects of the Mobile Age*. Springer-Verlag, London; pp. 46–61.

Laurier E (2001b) Why people say where they are during mobile phone calls. *Environment and Planning D: Society and Space*, **19**: 485–504.

Licoppe C (2002) Two modes of maintaining interpersonal relations through telephone: from the domestic to the mobile phone. In: Katz J (ed.). *Machines that Become us: The Social Context of Personal Communication Technology*. Transaction Publishers, New Jersey; pp. 171–185.

Ling R (2001) "We release them little by little": maturation and gender identity as seen in the use of mobile technology. *Personal and Ubiquitous Computing*, **5**(2): 123–136.

Ling R (2003) Fashion and vulgarity in the adoption of the mobile telephone among teens in Norway. In: Fortunati L, Katz J, Riccini R (eds). *Mediating the Human Body*. Lawrence Erlbaum Associates, New Jersey; pp. 93–102.

Ling R (2004) *The Mobile Connection. The Cell Phone's Impact on Society*. Morgan Kaufmann, San Francisco.

Ling R, Yttri B (2002) Hyper-coordination via mobile phones in Norway. In: Katz JE, Aakhus M (eds). *Perpetual Contact: Mobile Communication, Private Talk, Public Performance*. Cambridge University Press.

Lobet-Maris C (2003) Mobile phone tribes: youth and social identity. In: Fortunati L, Katz J, Riccini R (eds). *Mediating the Human Body*. Lawrence Erlbaum Associates, New Jersey; pp. 87–92.

Lobet-Maris C, Henin L (2002) Hablar sin comunicar o comunicar sin hablar: del GSM an SMS. *Revista de Estudios de Juventud*, **57**: 101–114.

Martin M (1988) Rulers of the wires? Women's contribution to the structure of means of communication. *Journal of Communication Inquiry*, **12**: 89–103.

Martin M (1991) *"Hello Central?" Gender, Technology and Culture in the Formation of the Telephone System*. McGill-Queen's University Press, Montreal.

Marvin C (1988) *When Old Technologies Were New. Thinking about Electronic Communication in the Late Nineteenth Century*. Oxford University Press.

Miles S, Moore K. (2004) Young people, dance and the sub-cultural consumption of drugs, Addiction.

Moyal A (1992) The gendered use of the telephone: an Australian case study. *Media, Culture and Society*, **14**: 51–72.

Murtagh GM (2001) Seeing the "rules": preliminary observations of action, interaction and mobile phone use. In: Brown B, Green N, Harper R (eds). *Wireless World. Social and Interactional Aspects of the Mobile Age*. Springer-Verlag, London; pp. 81–91.

O'Hara K, Perry M, Sellen A, Brown B (2001) Exploring the relationship between mobile phone and document activity during business travel. In: Brown B, Green N, Harper R (eds). *Wireless*

World. Social and Interactional Aspects of the Mobile Age. Springer-Verlag, London; pp. 180–194.

Oksman V, Rautiainen P (2002) "Perhaps it is a body part": how the mobile phone became an organic part of the everyday lives of Finnish children and teenagers. In: Katz J (ed.). *Machines that Become us: The Social Context of Personal Communication Technology.* Transaction Publishers, New Jersey; pp. 293–308.

Oksman V, Rautiainen P (2003) Extension of the hand: children's and teenagers' relationship with the mobile phone in Finland. In: Fortunati L, Katz J, Riccini R (eds). *Mediating the Human Body.* Lawrence Erlbaum Associates, New Jersey; pp. 103–112.

Perry Ch (1977) The British experience 1876–1912: the impact of the telephone during the years of delay. In: De Sola Pool I (ed.). *The Social Impact of the Telephone.* MIT Press, Cambridge, Massachusetts; pp. 69–96.

Rakow L (1991) *Gender on the Line: Women, the Telephone and Community Life.* University of Illinois Press.

Rheingold H (2002) *Smart Mobs. The Next Social Revolution.* Perseus Books Group, Cambridge, Massachusetts.

Rivière C (2002) La práctica del mini-mensaje en las interacciones cotidianas: una doble estrategia de exteriorización y de ocultación de la privacidad para mantener el vínculo social. *Revista de Estudios de Juventud,* **57**: 125–137.

Roos JP (1993) Sociology of cellular telephone: the Nordic model: 300,000 yuppies? Mobile phones in Finland. *Telecommunications policy,* **17**(6), www.valt.helsinki.fi/staff/jproos/mobiletel.htm

Sacher H, Loudon G, Pellijeff O (2001) Creating a third generation wireless application for teenagers. In: *Mobilize! Interventions in the Social, Cultural and Interactional Analysis of Mobility, Ubiquity and Information and Communication Technology, Conference Proceedings.* DWRC, University of Surrey, Guilford.

Sennett R (1986) *The Fall of Public Man.* Faber, London.

Sherry J (2001) Running and grimacing: the struggle for balance in mobile work. In: Brown B, Green N, Harper R (eds). *Wireless World. Social and Interactional Aspects of the Mobile Age.* Springer-Verlag, London; pp. 108–120.

Skog B (2002) Mobiles and the Norwegian teen: identity, gender and class. In: Katz JE, Aakhus M (eds). *Perpetual Contact: Mobile Communication, Private Talk, Public Performance.* Cambridge University Press.

Taylor A, Harper R (2002) *Age-Old Practices in the "New World": A Study of Gift-Giving Between Teenage Mobile Phone Users,* CHI 2002, 20–25 April, Minneapolis, USA.

Taylor A, Harper R (2003) The gift of the gab: a design oriented sociology of young people's use of mobiles. *Journal of Computer Supported Cooperative Work,* **12**(3): 267–296.

Vincent J (2003) Emotion and mobile phones. In: Nyíri K (ed.). *Mobile Democracy. Essays on Society, Self and Politics.* Passagen Verlag, Vienna; pp. 215–224.

Vincent J (2004) Are mobile phones changing people? *Proceedings of the Fifth Wireless World Conference.* DWRC, University of Surrey, Guilford.

Weilenmann A (2003) Doing mobility, Doctoral dissertation, Department of Informatics, Goteborg University.

Weilenmann A, Larsson C (2001) Local use and sharing of mobile phones. In: Brown B, Green N, Harper R (eds). *Wireless World. Social and Interactional Aspects of the Mobile Age.* Springer-Verlag, London; pp. 92–107.

Willey M, Rice S (1933) *Communication Agencies and Social Life.* McGraw-Hill, New York.

Wurtzel AH, Turner C (1977) Latent functions of the telephone: what missing the extension means. In: De Sola Pool I (ed.). *The Social Impact of the Telephone.* MIT Press, Cambridge, Massachusetts; pp. 246–261.

Young P (1991) Person to person. In: Granta (ed.). *The International Impact of Telephone.* Cambridge.

Kids will be Kids: The Role of Mobiles in Teenage Life

3

Richard Harper and Lynne Hamill

3.1 Introduction

Mobile phones and Short Messaging Services (SMS) (or "text") in particular are new social phenomena, much marvelled at and much commented upon (e.g. Katz and Aakhus, 2001; Brown *et al.*, 2002; Ling, 2004; Harper *et al.*, forthcoming). This success is said to be because mobiles allow new levels of micromanagement in an age of fraught and tight deadlines (Plant, 2002) or because they allow communities to create and sustain their own language networks (Sanda, 2003: 71–81). But mobiles are often criticised because they are a distraction from true engagement with people at a face-to-face level (as discussed by Reid and Reid in Chapter 6) and because, along with other technologies, they will dissolve the "civic" society (for a sample of articles on these topics see Nyiri, 2003).

What is certain is that mobile communication, whether it be fully duplex telephony or SMS traffic, is on one hand "merely" people communicating, undertaking the prosaic activity of chitchat within the frame of a particular medium, yet at the same time, many other things too. Talking is after all not always merely chitchat: it is made up of very many different goals, functions and content. Indeed it can be argued that texting is a microcosm of society at large, as Harvey Sack's taught long ago in relation to everyday conversations (Sacks, 1992). The question that bedevils research in this area is not whether texting, say, is an expression of a society that has not changed, nor whether it is a manifestation of something new. The problem is that society tends to immerse new technologies in ways that make the uniqueness of the new technology – whatever those unique properties might be – less important, less visible than might be wished: as a result, whether there are changes or not is all too difficult to see.

Consider teenagers: they have been quick to adopt mobiles. By 2002, teenagers aged 14–18 years in the UK were as likely as adults to own a mobile: around two-thirds. Even among those aged 11–14 years, over half had one (ONS, 2004). Therefore, it is not surprising that much has been written about teenagers and their use of mobiles. As noted above, sometimes it is claimed that this technology is enabling new practices amongst teenagers and sometimes old ones in a new form; sometimes practices that are good for teenagers and sometimes ones that are less desirable (at least from some peoples' point of view, if not teenagers' themselves).

It may be that there can be no answer to the question of whether mobiles are good or bad, or whether teenagers are being changed by them or are changing them; it may be simply too early to tell. It may be, in any case, that such contrastive questions – good/bad, altering/altered by, enabling/constricting and so on – are simply too crude to capture the full richness and diversity of human practice, teenagers having their own variant of these practices. Perhaps mobiles do create change in some aspects, perhaps they produce stasis elsewhere.

It could be that a better approach to social inquiry is not to posture around these polar opposites, but to camp instead in areas where certain salient aspects of particular realms of social life are clear to see and remark upon: this less burdened approach to inquiry might produce more carefully honed insights that concerns with grander claims make difficult to attain.

Our own research on teenagers, for example, has shown that there are many ways in which mobiles are enabling teenagers to do what they have always done and which they have hitherto had less support to achieve; and has also provided examples of where they are using technology to undertake tasks that in previous situations they could not have imagined. Yet to confuse matters when some of these apparently new practices are examined over a longer time frame (and thus not with regard to their immediate impact or shaping) and instead as regards how they are adopted and settled on with the passing of time, the patterns in question turn out to be less distinct, less new than was thought. Indeed they sometimes begin to appear to be redolent of inertia rather change, stasis rather than movement. Thus although mobiles might be new and although what teenagers might do with them might seem to be somehow different, these same practices seem to be slowly pulled into a world that will not change in quite the ways teenagers and the commentators want or even imagine.

Now it is probably too early to say whether this interplay of the persistent and the novel, the habitual and the yearned for, will turn out in favour of one side or another, or even whether such inquiries are worth pursuing. But this chapter offers some evidence as to how particular aspects of what mobile telephony enables is manifesting in particular places and particular times. These may be snapshots of processes that are endemically historical, slow and/or complex, but nonetheless, we believe, worth considering.

Two issues will be focused on. By selecting just two it should be clear that every dimension of the "teenage-mobile axis" is not being explored; that would be too ambitious, even if possible. These two are selected because of their salience in our data and because they illustrate the importance of matters which seem to have been somewhat conspicuously left out by much of the research on teenagers. Whatever the reason for this absence, these two topics have much merit on their own and in many ways are sufficiently different to make their juxtaposition worth attempting.

They are, first of all, to do with the all too social process of cost management, which is a somewhat oblique way of inquiring in to the patterned ways in which mobile phones are paid for and how these costs are negotiated. The financial management of mobile telephony by teenagers is not something that is solely the concern of these teenagers, nor is it a cost that stands alone among all the costs that they are, or are thought by others to be, responsible for. Rather the costs of mobile devices stand as testament for the management of a host of costs and this management is chronically visible to others who might have some interest (real or imagined) in those costs. And here lies the rub: in being a visible philosophy, in being conspicuous as either the costs of a spendthrift or a thrifty person, as either the costs of one who is reasonable or wanton, and in being so costs that are measures of other behaviours that incur costs, then these costs become the topic of extensive and chronic social exchange between teenagers and those with whom they share their lives.

The second topic relates to what might be called conversational turn-taking systems, including text communications. Systems of etiquette and propriety govern mobile communication as indeed they govern all forms of communication. This section looks at how these systems are being deployed and manifested with this new technology, presenting some evidence hinting at how these systems, old, well practised, known yet oddly unremarked in everyday life except at their breaking, are being used to create fine but often consequential distinctions between teenagers of different ages, gender and social connection. Further, the evidence is suggesting that it is individuals in their late teen years who seem most rigid and elaborate in the way they impose these systems, excluding some from communicating and admitting others strictly in accordance to certain rules of access that older and younger age groups worry less about.

These two topics – the question of who pays what and when on the one hand and who calls who, when and about what on the other – do not by any means conclude all that could be said about mobiles and teenage life but they do point towards some directions of understanding, towards a sensibility for understanding the "just what" of the mobile age.

3.2 The Cost of Mobile Phones Versus Costs in General

As part of an ongoing series of projects with a major mobile network operator in the UK, data was collected on the evolution of fixed and mobile phones in family life. In one project, diary studies and interviews were undertaken in households in the UK and Sweden. Some 59 persons from some 21 households where interviewed and monitored. All had both mobile and fixed lines available to them. The research focused on the grounds for the use of either.

One aspect of teenage life studied was how teenagers dealt with mobile costs and this in part had to do with how they learnt that spending on mobiles would be at the cost of spending on other things, what economists call the "opportunity cost". It is a concept that is central to economics which is, after all, about the allocation of scarce or limited resources. For teenagers this means that if they spend their money on, say, a top-up card for their mobile, they will not be able to go out on the forthcoming Saturday night, or if they do, they will not be able to afford to drink. In other words, they are learning that it is not possible to have everything that they might want, an important lesson as they grow from children into adults. Now, clearly, though opportunity costs may be made up of a host of aggregated factors, contingent, variable and/or complex, and though mobile devices may have properties that make these costs even more complex – the existence of the virtual address book on a mobile and its absence on fixed line devices occluding financial judgement of utility of each for example – it was clear from the outset of our research that the cost of mobiles was and is measured against other kinds of costs. But this measurement was and is not just about relative value, the opportunity cost mentioned, but the costs of mobiles was and is also to be understood in terms of the philosophy that a particular teenager uses for the management of dealing with costs in general. And what this, in turn entailed, is how the question of costs no longer remains in the hands of the teenager but becomes somehow – and this somehow is of course the issue at hand – a question for other people too. Now given that these are teenagers, these other people are not just anybody from any time and place; it is mum and dad. And not just any old mum and dad but this particular Mum and Dad and this particular Teenager. This in turn means that the foibles, moralities and – using a term that might give exaggerated dignity to what is an often impulsive and ill-considered matter – the philosophy of those parents will be brought to bear in conflicting ways on the philosophy of the teenager in question. What starts out as a question of money ends up being about very much more indeed.

One interview with a father in a UK household conveys the gist of this take on the economics of the mobile. He said, "You know mobile phone bills are about the only thing I can talk to my daughters about when what I really want to talk to them about is not eating things out of the fridge and not telling anyone. I mean, they have got to learn that there are other people in the house and the only way I can think of making them do this is by having a talk about mobile phone bills and then I can talk to them about money and living together and sharing things without coming across as pompous, like some Victorian patriarch".

Another example from the same family illustrates this in another way. When the mobile phone bills – or direct debit statements to be precise – arrived, this dad would pick them up, open them and leave them around for his two girls, late teenagers, to come across. He would put them on the kitchen table or on the fridge so that he could guarantee they would see them. They would thus not only be aware of their existence but would be also aware that "Dad had put them there since he wants to lecture us".

He was not, however, concerned with the size of the phone bills. As he put it, "That's up to them". His concern was to discuss how the respective phone bills identified certain behaviours which he viewed as irresponsible; these costs may have had nothing to do with the mobile phones themselves. As a case in point, he had noted that when the girls were both at home (one had just started at University and had been home for two holiday periods), his own fixed line phone bill went up substantially. His phone statement showed that this was primarily because of calls to mobile phones rather than to other fixed lines. Now, his concern was that for many of these calls it would have been cheaper had they been made from a mobile on the same network. He believed also that in many cases the girls' own phones were on the right networks for this; and that it was the girls who chose to make the calls on the fixed line not simply or even partly because they knew their dad was paying, but because they could not be bothered to find their own phone. Their costly behaviour was simply irresponsible behaviour.

The reason he wanted to talk with the girls, then, was that he did not necessarily mind paying bills, including their own, but he did mind paying bills unnecessarily. Bills could be reduced if individuals thought about the overall economy of the family. For him the issue was that the girls treated expenditure as a primarily individual rather than a collaborative matter and therefore they did not act in a way that reflected concern for others. In crude terms, if the girls recognised that some costs were shared then he believed that their behaviour would be different. Their use of the fixed line phone when a mobile would have been cheaper would have been an instance of this. By addressing this behaviour he hoped that the girls would adjust their behaviour for all shared matters in the house, whatever it might be. To be able to conduct oneself with respect to others was a matter vastly more important than the actual costs of something in particular. It was, if you like, a question of morality.

Now, it is worth noting that there is nothing new in the existence of difficulties between parents and teenagers when it comes to the management of phone costs be they mobile or fixed line telephony. Looking back to a study undertaken in the mid-1990s, before many teenagers in the UK had mobiles, it was found that "Nearly two-thirds (65%) of British 14–17-year olds received complaints about cost" (Haddon, 1998). Furthermore, there is nothing unusual about parents and teenagers entering in to discord about costs in general. This behaviour is a normal part of family life. What is perhaps more interesting (in being less often remarked upon in the scientific literature) is how technologies of various sorts become the indirect pretext of such happenings. Our own prior research on the use of paper letters, for example, found that one of the (social) organising consequences of paper bills is a pretext for such arguments (Harper *et al.*, 2000). Paper bills are put in places that are likely to ensure their noticing and this in turn ensures a kind of discussion, an awareness of the bill that can be brought up in conversation over breakfast, dinner or tea. Similarly with the mobile phone bill its arrival, its judicious placement in a place where all can see, becomes the pretext for a parent–teenager "moment".

Bills, particularly paper ones, their arrival and their contents are then of curious importance to teenagers and those they share their homes and lives with; but just how this manifests itself is sometimes remarkably attenuated. For example in the same household mentioned above, the father recounted the following story: "Look this sounds daft but I had some sausages in the fridge to make dinner and when I went to the fridge I found that (one of his daughters) had eaten them, well at least it must have been her. Now, they are only sausages – though they were special ones I had bought – and I don't mind them eating them but now there isn't anything to cook and I don't want to go up to (the nearest supermarket)."

In the first example the father did not worry too much about mobile phone bills but their arrival was the only pretext he could think of that would enable him to get the girls to sit down over dinner and have, as he put it, "a rational conversation" about learning to share. In the instance he is reporting here, his real agenda was about the sausages, but he felt that the issue of sausages *per se* would be simply laughed at by his girls. He was probably right. Yet, only through addressing a matter that they thought was potentially serious, namely phone bills, could he indirectly address matters that they thought were inconsequential but he thought symbolic. In short, he wanted to use conversations regarding mobile phone bills to raise the possibility that they might start behaving in different ways with regard to other matters.

One might put this in a larger context: when these teenagers had been children, they might have simply taken without asking and used without commenting; as they were getting older and, presumably, as leaving home became increasingly imminent, he wanted them to start living in a manner where shared responsibility was the norm. His view was that part of moving on from being a teenager has to do with the ability to take on responsibility. One of these responsibilities is for household bills; another has to do with consumption of shared goods, like groceries. The girls should cease behaving with little or no concern for others in the same space; they should start considering how their own behaviours would affect others. In a phrase, he simply wanted his girls to start behaving like adults: recognising that if the fridge was stripped of food then others in the house might be left hungry by the end of day, having planned to eat that same food.

3.3 The Social Etiquette of Calling

The rub of the matter, then, for these teenagers and the household that they are part of is the difficult, socially organised process of movement from one social role to another. Here a father is trying to facilitate that in the best way he can; though doubtless his daughters thought his efforts at best harmless, at worse tiresome; almost certainly his actions caused them to giggle. But this change in social status is not solely achieved through the coercion and benign encouragement of others; teenagers also contribute to it themselves, albeit that the way they do so, and the manifest consequence of this achievement, may not be so visible to themselves.

This can be illustrated by looking at the issue of turn-taking on mobile communications, texting being one genre of these communications. In part this argument is also about the question of the social shift in the competence of teenagers, but as with our prior argument, our concern is not with the shift in status itself, but how that shift, along with many other changes in teenager behaviour, turn out to include processes that somehow draw teenagers and the technologies they use in to the steady, persistent, innocuous yet coercive patterns of "how things ought to be". In the first case, as regards costs, it was parents trying to coerce teenagers to act like them; now, in this second set of examples, it is teenagers coercing each other to behave in ways that are, we want to suggest, more traditional, more conservative (even) than might be expected. The coercions are those to do with the rituals of who can say hello to who, when and why. Instead of being machines that open up the possibility of a communications free for all, teenagers make mobile technologies the pretext for making the rules and conventions of social interaction ever more rigid and hierarchical.

Approaching this from another place, where teenage practice may be somewhat distinct, gives some contrastive colour to our own evidence. Ito (2003) reports that one of the skills that Japanese teenagers learn, with their mobiles, is to use spatial matters as a resource for managing the social etiquette of communication. She cites various examples of how teenagers will end a mobile phone call on the bus when that bus is about to reach the stop they want to use; they tell a friend to stop texting when they are about to enter a class. Now, the important issue here is that in the instances she cites the possibility that these relationships between space and action are contingent, and arbitrarily appropriated comes to mind. That is to say, perhaps the teenagers were not saying to their friends "Oh the bus stop is near!" because it really was near; it was rather that they were using the bus stop as an excuse to manage the call. After all, one of the properties of mobile communications is that physical and spatial matters that might impact upon the management of conversation are not equally distributed. Thus the caller may have no knowledge as to whether the person they are calling is indeed about to enter a class room or get off a bus. Of course it is certainly true that the caller may have some information: the noise of traffic, for example, may indicate that the person called is indeed on a bus; the screeching and bellowing of kids may suggest that the one they are calling is loitering outside a class room. But these resources are at once an indication, auditory clues if you like, that the caller may invoke: "Oh are you about to go into class?" one can hear them say. But they are ultimately the tools of the one who is in that location: it is only they who can deny that it is the sound of kids and say perhaps that their television is on; it is only they who can say that they are walking down the street when in fact they are on a bus. They cannot dissemble too aggressively, needless to say, since that would be bring in to doubt their accounts; the point is that this imbalance of what one might call local knowledge (or situated knowledge) is such that it provides an easy resource for managing the process of calling.

Why is this resource needed? Why do they need to "manage calls"? This almost sounds like adults worrying about saying the "right thing", hardly a concern that one would imagine teenagers to fret about. Yet what Ito suggests, and indeed much of the other research on teenage life confirms, is that teenagers do indeed worry about this although this worrying is graded and structured according to age.

Crudely speaking, new users do not know how to manage mobile calls at all and this results in them using the phones excessively; it is only gradually as they move from early to later teens that these skills become more astute and refined. These skills have many forms and their evolution is itself a measure of the general social skills of the individual in question. In Kaseniemi's "Mobile Message" (2001) for example, Finnish teenagers report how tiresome they find friends who have just got their first mobile: apparently they phone and text all the time. Once they have got over this excitement they start to use the devices more "appropriately", it is reported.

What this means is itself variable and complex. The same set of subjects report differences in the behaviour of the two genders: girls treat what they share and exchange over the mobile as more private than boys. So girls modulate what they say according to the gender of the person they are calling. Taylor and Harper (2003) are among others who have noted that there are ritual communications that need to be undertaken when girls and boys are going out together: the goodnight text sent from a boy to a girl last thing at night is now a social requirement, for example. Failure to deliver the message results in a summons the following morning in the playground. Sending a steady stream of little notes throughout the school day is also a measure of devotion and adoration; the absence of the same an indication that an "item" (an idiomatic label for a couple) is not what they once were. All these little differences, in content, in the frequency of calls, in who is calling who and so on, are not only visible to those involved and merely matters of private moments; they are also matters of public interest since all are subject to the same patterns, exchanges and rituals. Boys complain to other boys about the oppressive need to send goodnight texts; girls about the slovenly failure of the boys to send them, and so forth.

These patterns are of course somewhat varied with different codes being applicable in different societies and cultures. A comparison between Japan and France, for example, highlights curious differences: curious since they are suggestive that the simple views often deployed when thinking about these two very different places and cultures have some merit, albeit that these views are treated, even though used, as crude. Riviere and Liccoppe (forthcoming) report that texting is used between persons of different social status so as to avoid the faux pas of interruption; between intimates such as husband and wife, no such fear is present and thus voice calls are made any time day or night. In contrast, in France, texting is used not so much to avoid the problem of interruption as to avoid the possibility of emotional violence that goes with close relationships: thus girls would prefer to text their complaints to a boyfriend since this would not result in a physical outburst from that same boy; the boys prefer to text their own concerns since the girls do not respond with tears and weeping. Somehow text not only avoids these all too real physical reactions being seen, they also make them less likely to happen. Girls apparently find themselves less weepy when they communicate with texts; boys less prone to violence.

These examples illustrate, then, that mobile technology is affording more refined and controlled patterns of behaviour as regards who says what, to whom and when; a finessed approach to the all too human task of talk. It also shows that teenagers around the world, in countries and places that are in many ways quite dissimilar, are beginning to develop fairly common, yet elaborate patterns for mobile-enabled communication amongst themselves. It would appear also that these patterns slowly mature as teenagers get older: what was accepted when 13 years old is laughed at and a source of embarrassment by the time they are 18 years. There is in other words, a self-accomplished sophistication amongst teenagers, a sophistication as regards the who, the when and the what of mobile connectivity.

It is now time to come back to our own zone, our own evidence about these patternings and codes. Turning again to our studies of home life one of the things that came out of our research was that this sophistication is not only with regard to the calls made between teenagers, but also relates to the problem of how others, particularly parents, might interfere with these calls and attendant rules of propriety. It appears that one of the reasons teenagers in the UK and in Sweden like to use the mobile when calling from their home, and one of the reasons why they like to call a mobile rather than a fixed line, is that thereby they can guarantee who they will end up talking to. On the one hand, the receiver of a call can see the name of the caller, presented through the functioning of the virtual address book entry associated with that number. This follows on from what has been reported earlier; but in addition the other side of the arrangement here has to do with socially invested practice that holds that only the owner of a mobile will pick it up. Thus, a caller knows that their identity will be displayed at the other end of their call; but they also know who will answer. The mobile phone has come to be one of those articles that remains essentially one person's sole responsibility. Thus a call to that person's mobile will rarely be answered by someone else, but only by that person. By contrast a call to a fixed line could summon anyone within the space in which that fixed line phone rings.

There are many things that could be noted on this binding of technological functioning and socially arranged investment of value; the value here being one of ownership and rights to access, control, even touching of a device. For now though the focus is on how it would appear from this evidence that teenagers in particular are loath to speak with their friends' families, particularly parents. "Oh they are so awkward", as one of our respondents remarked. Apparently teenager–parent conversations, whether they be within a family or across families, are always difficult. There is, needless to say, a reverse to this too. Sometimes a fixed line is used to communicate to the home so that the caller can discover who is there, so as to open up a serendipitous conversation with mum, dad or their brother.

How often or frequent this is, however, is altogether another matter. Though this might be a tactic that is available, this is far from saying that it is a tactic used often, whether it be daily or more likely at weekends. In three households, for example, daughters explained that they deliberately choose to call home on the landline so as to avoid constraining who they talked to. Alas, even as these three daughters expressed this, their mothers, being interviewed at the same time, contradicted them. As one mother put it, "You say that but you never do. You don't want to talk to your dad so you always call me on the mobile and then dad has to try and catch you later. You know he never does."

Whatever one might say about how members of families express their intentions and yet manage to act in ways that obscure those same intentions, and similarly whatever might be said about the oft-noted strains of father–daughter relationships in our study of home life in the UK and Sweden, key to all of the above is one property of mobile technologies and that is how the virtual address book, when combined with the assumption that only one person has rights to answer a mobile phone, creates what might be called a tight coupling of social systems of propriety and technology (Berg *et al.*, 2003).

This tight coupling has all sorts of ramifications and deployments. In other research we have undertaken, in this case teenage life in schools southwest of London, the virtual address book was found to manage, in highly demonstrative ways, the alteration of rights to call and reject a call. When these teenagers no longer wanted to communicate with another, as in the case of a girl breaking up with a boy, the name of the person out of favour would be ceremonially, one might almost say ritually, deleted from the address book. "I deleted him" was a phrase that was often used. Such deleting of data entries did not inhibit the person in question calling (the one who had been removed from civil discourse) but it did mean that when the call is made no name comes up on the screen: it is the summons of an anonymous person.

The long and short of it is that our evidence of teenage behaviour, as well as evidence others in different places have gathered, is showing how certain properties of mobile devices are being leveraged to effect patterns of communication. Though it might seem natural to answer any and every call whoever makes it, and indeed this has been habitually the case with users of fixed line telephony, with mobiles and their technological infrastructure that provide caller line identification, when combined with the socially sanctioned rights of ownership and control, the phones themselves are now being used to let the recipients of a call determine whether they wish to answer or not. They do so by grading their decisions according to social rights. People, or in this case, teenagers, who have the right to call have this right embedded in their "presence" in the address book: those who do not have this right have it demonstrated to them by their exclusion from the address book. Membership and exclusion is not permanent. It is in some ultimate sense flexible: dependent in one way or another upon the state of relationship between two persons. But when exclusion occurs it is indeed a heavy sanction, shifting the social status of an individual from the in-crowd to the out-crowd; from one who matters to one who does not, from one whose call will be answered to one whose call will provoke silence.

Teenagers then, especially as they grow towards late teens, use a link between the virtual and the real to manage the details of their phone communications. They appear to do this in ways that can be described as rigid. Teenagers in the schools we studied really did avoid answering calls that do not have a caller ID; similarly with the teenagers in our home study. The result of this is that instead of being available to contact by anyone at any time, one purported reason why mobile technology was devised, the practices of teenagers result in constraints being placed on their social worlds: these ensure that only those who have a right to contact them do indeed contact them and those that are excluded rarely make the effort and if they do will be rebuffed.

In summary, what we are finding is that mobile technologies are allowing teenagers to work at their relationships; that this working may entail communicating more frequently; and that it is allowing teenagers to effectively direct their communications to the ones they want. We are seeing also that they can use the technology to embody what might have hitherto been unfilled thoughts, ideas and ambitions about who can and cannot contact them. The result is a system of social etiquette that is at once complex, subtle, highly graded and punitive: this is the work not of those who have power over teenagers, it is they themselves who create these tongue-tied processes. Others have noted this too: to extend the list of researchers mentioned above, similar evidence has also been uncovered in Finland (Kopomaa, 2000), Norway (Bakken, forthcoming), Germany (Hoflich and Gebhart, 2003) and the Philippines (Ellwood-Clayton, 2003).

3.4 An Historic Perspective

The question that arises from all of this is, then, how many of these are new and how many are old? What role does the newly adopted technology play in the balance of these alternatives? In the pull towards financial responsibility are teenagers slowly adopting what they have always been taught, long before the invasion of the mobile device and the direct debit? Are the ways in which they limit and restrain their social worlds to those who are "in" as opposed to those who are "out", the kinds of rigidities that have existed before?

Turning the clock back a generation to 1967 – when The Beatles were singing "All You Need Is Love", film audiences were first enjoying The Graduate and when sociology took a very different approach to exploring its subject matter – then despite all these differences in the concerns of the time, in the methods used to explore those concerns and even in the technologies being reported, there are some stark possibilities that things are not so different. Schwartz and Merten (1967), for instance, when investigating the social life of teenagers in the USA, tried to delineate what made certain kids "cool" and others uncool. The measures used by teenagers to distinguish between these were different to those used by the teenagers we studied, just as the approach used by us and our sociological forbears in the analytical work differs: now there is much less concern with socio-structural fit and internally consistent ideologies, for example, much more with the "lived experience" of particular places and times. But all these differences notwithstanding, what Schwartz and Merten found was that teenagers created rigid hierarchies amongst themselves, hierarchies that evolved with age. They report: "One informant described what happened to a fraternity which did not make the shift to 'socie' patterns: The Lambdas aren't well liked now because the Lambdas don't drink, and the other kids (the cool ones) are all getting to drink, and they [the Lambdas] are not well liked anymore because they look down upon it [drinking]. So now if you want social prestige with the kids you wouldn't dare mention the Lambdas."

This could be altered to aver to mobile phones, with the Lambdas remarking that they are still calling everyone they know because they can, whilst others, being "cooler" and more popular, know how to limit their calls to only those who are "in". Thus the issues that distinguish social groups are not now the use of intoxicants, as it was in the 1960s, but the management of social contact: a different aspect of teenage life, but one that teenagers probably find more affecting.

It may be that the Lambdas and those who achieved social status through drinking would find today's teenagers different not only in dress and language but in the major drama they make of such small things as address books, and the rhythms and rights of telephony; indeed, they may find today's teenagers unmanly and unfeminine in equal part. But these are in a sense superficial details and our concern is rather whether today's world is different from theirs in any measurable or salient way. Teenagers did then and do now create their own social order, their own hierarchy. What we have seen is that the prosody of calling and answering, of content and topic management with mobiles, has become an increasingly artful practice for today's teenagers; and that after a certain age, an inability to manage these issues gracefully is viewed as a measure of immaturity. In other words, as they grow, teenagers themselves start behaving in ways that distinguishes those who are becoming adult and those who are not. These skills and competences have to do with the social rituals of when to address someone, how to address someone and what to say. These are at once ornate yet every day, prosaic yet artful. They are about the socially achieved skills of ensuring the appropriate intersections of time, place, content and persons.

This intersection is not the only one that teenagers have to deal with. The first empirical section of this chapter showed how parents used mobile phone bills to educate teenagers about economising and sharing, to create an appropriate fit between what is at hand, who is around and what needs to be done. Is this new? Long before the mobile phone, parents have presumably tried to inculcate their offspring with their own ideas about frugality and consideration for others; to instill a sense of the moral order of the household. Who can say? What is certain is that the introduction of a new technology, mobile devices, has certainly played into these larger processes and patternings. It has done so in ways that sometimes creates shifts forward and sometimes shifts backward; that sometime forces movement to new social practices and sometimes changes towards an idea of how things ought to be and perhaps were, once, in some distant time before the mobile phone rang and someone was overheard to say "Can you talk now? Where are you?"

References

Bakken F (forthcoming) SMS use among deaf teens and young adults in Norway. In: Harper *et al.* (eds). *Inside Text: Social, Cultural and Design Perspectives on SMS*. Kluwer, Amsterdam.

Brown B, Green N, Harper R (eds) (2002) *Wireless World: Social and Interactional Aspects of the Mobile Age*. Springer, London.

Ellwood-Clayton B (2003) Virtual strangers: young love and texting in the Filipino Archipelego of Cyberspace. In: Nyiri K (ed.). *Mobile Democracy*. Passengen Verlag, Vienna; pp. 225–239.

Haddon L (1998) Il Controllo della comunicazione. Imposizione di limiti all'uso del telefono. In: Fortunati L (ed.). *Telecomunicando in Europa*. Franco Angeli, Milano; pp. 195–247.

Harper R, Evergeti V, Hamill L, Moray N, Watson D (2000) *The Future of Paper-mail in the Digital Age: An Investigation into the Affordances of Paper-mail*. Digital World Research Centre, University of Surrey, Surrey.

Harper R, Palen L, Taylor A (eds) (forthcoming) *Inside Text: Social, Cultural and Design Perspectives on SMS*. Kluwer, Amsterdam.

Hoflich J, Gebhart J (2003) *Vermittlungskulturen im Wandel: Brief–E-mail–SMS*. Lang, Frankfurt am Main.

Ito M (2003) Mobile phones, Japanese youth, and the replacement of social contact. In: *Proceedings of Frontstage–Backstage: Mobile Communication and the Renegotiation of the Public Sphere*. Ling R, Pederson (eds). Telenor, Norway, Summer.

Kaseniemi E (2001) *Mobile Message*. Tampere University Press, Tampere, Finland.

Katz J, Aakhus M (2001) *Perpetual contact: Mobile Communication, Private Talk, Public Performance*. Cambridge University Press, Cambridge and New York.

Kopomaa T (2000) *The city in your pocket: Birth of the Mobile Information Society*. Helsinki, Gaudeamus.

Ling R (2004) *The Mobile Connection: The Cell Phone's Impact on Society*. Morgan Kaufmann, New York.

Nyiri K (ed.) (2003) *Mobile Democracy*. Passengen Verlag, Vienna.

ONS (2004) *Social Trends 34*. Stationery Office, London; pp. 203–204.

Plant S (2002) *On the mobile: the effects of mobile telephones on social and individual life*. Report commissioned by Motorola.

Riviere CA, Liccoppe C (forthcoming) France/Japan: major trends with respect to mobile phones. In: Harper R, Palen L, Taylor A (eds). *Inside Text: Social, Cultural and Design Perspectives on SMS*. Kluwer, Amsterdam.

Sacks H (1992) In: Jefferson G (ed.). *Lectures on Conversation*. Blackwell, Oxford.

Sanda K (2003) The fall of linguistic aristocratism. In: Nyiri K (ed.). *Mobile Communication*. Passengen Verlag, Vienna; pp. 71–81.

Schwartz G, Merten D (1967) The language of adolescence: an anthropological approach to the youth culture. *American Journal of Sociology*, **72**(5): 452–468 (available from www.jstor.org).

Taylor A, Harper R (2003) The gift of the gab: a design oriented sociology of young people's use of mobile. In: *CSCW: An International Journal*, Kluwer, Amsterdam; pp. 267–296.

An SMS History

4

Alex S. Taylor and Jane Vincent

4.1 Introduction

This chapter examines how the seemingly unsophisticated messaging technology, Short Messaging Service (SMS), has found itself centre stage in contemporary social life. It examines how the position of SMS has been conflated, intertwining the social, economic and technological aspects of the capability, transforming the messaging service into something that is much more than merely a primitive means of composing, sending and receiving the alphanumerical messages known as "texts".

Tracing certain elements of the SMS history since the early 1990s, the chapter explores how the combination of business and technological developments, like the shift to interoperability between networks and the capacity for a flat rate charging model for message delivery (rather than the previous paging model), precipitated a swell in the popularity of SMS. Factors such as these are seen alongside particular social developments (e.g. the early uptake of SMS among young people) and how they worked collectively to provide sufficient impetus for the widespread uptake of SMS. This chapter also examines how technological constraints, such as the 160-character limit on messages, the limitations of the alphanumeric character set and the design of mobile devices were closely linked to the emergence of an SMS "shorthand".

Unpacking these emergent properties, these transformations, lays bare the complicated interrelations that subsume technology-in-progress. The intention is to reveal that no simple path can be drawn to explain the developments in and uptake of technologies (e.g. de Laet and Mol, 2000; Callon and Rabeharisoa, 2003). Through the chosen examples, we mean to contest the commonly held assumption that technology can be viewed as removed and somehow separate from society; that a neatly carved division exists between things and people, non-humans and humans (Latour, 1993). Instead, a picture emerges of SMS as yet one more collective that binds the worlds of non-humans and humans inextricably together. The chapter concludes with a discussion on how this examination of the SMS history has implications for reflecting on the next wave of developments in the broader mobile telephony project. Specifically, thought is given to the potential for picture messaging and what lessons might be learnt in order to better understand the progress and adoption of its specific feature set.

4.2 Talking "Texts"

On 3 March 2003, a national broadsheet newspaper in the UK published a short article about a 13-year-old girl who had written an essay in SMS shorthand (*The Daily Telegraph*, 2003). The article describes how the girl used the abbreviations and "hieroglyphics" commonly associated with texting to compose an assignment she had been given at her State secondary school in West Scotland. The article goes onto raise the looming spectre of falling standards in literacy amongst students in the UK and enlists a number of sources to examine this issue.

Rather than debate the moral panic that has ensued alongside the massive uptake of SMS amongst youth in UK and Europe, the motivation behind this chapter is to examine just how SMS has become talked about and even derided for its role in shaping social relations, literacy, privacy, crime and so on. Broadly, the aim is to examine how it is that a relatively simple messaging technology originally designed to be part of the maintenance layer of the GSM infrastructure has transformed into a messaging product and produced a new argot much discussed amongst the popular press, industry bodies and research institutions worldwide.

To paint this picture, attention is limited to certain aspects of the SMS history. Attention is given to those elements that directly relate to the uptake of SMS as a technologically and economically viable solution, and that have contributed to the service's considerable demand amongst mobile phone users. Although a complete account of mobile telephony and its tightly coupled sibling SMS might be considered ideal, the aim has been to trace the relationship between particular events and bodies to illustrate the complex interplay that must take place to get something out of the engineering laboratories, onto the shelves and into the hands of willing consumers. Thus, the story is not an attempt at a rendition of all the facts, but rather it serves as a timely reminder that the distinctions drawn between technology, society and business are blurred. It also stands to illustrate how an examination of these merged relationships is necessary in order to learn from the uptake and success of a technology such as SMS.

4.3 Technological Developments

The serendipity that would appear to have determined the present day texting phenomenon found its footing during the design of the

GSM Technology. GSM was originally the acronym for the Groupe Speciale Mobile, the working group from the European Conference of Postal and Telecommunications Administrations (CEPT) that laid out the specifications and standards for the first digital cellular system designed to enable the same mobile phone to be used on any network with GSM equipment. GSM was adopted as the commercial name for the service using the descriptive Global System for Mobile Communications to explain the acronym.

In the process of the network design and development in the 1980s, it was recognised that it would be possible to send short data messages at the same time as speech using what is known as the "signalling channel" or "layer" of the network. This channel, used in digital fixed telecommunications networks to monitor and check on the network, was not needed in the same way for GSM and thus offered spare capacity for the delivery and receipt of non-voice and, specifically, alphanumeric text-based data.

It was the development of the unified GSM standard in Europe that provided the technological infrastructure to support the general availability of this form of data exchange. Indeed as Kasesniemi (2003: 94) asserts in her study of text messaging in Finnish society "only the spread of GSM technology enabled the birth of the text messaging culture, which could not have emerged without the technological innovation of the SMS that was originally designed for an entirely different purpose". GSM differs from previous analogue cellular radio technologies in that the "air interface" – the part of the mobile phone communication that occurs between the mobile phone and the base station – is digital and the telecommunications standards that specify the service are largely non-proprietary. In other words, the basic services are the same regardless of manufacturer, thus enabling any GSM designed mobile phones to work on any GSM network. This was a major transformation for the mobile communications industry that had been offering national cellular systems using a variety of standards, sometimes with incompatible technologies competing in the same country. These services offered voice messaging and were sometimes sold as a package with radio paging services that provided a one-way alphanumeric messaging service in areas where voice coverage was not available. The aim of the common GSM standard was to provide a pan-European infrastructure with the ability to use the same mobile phone in any location where GSM coverage was provided. This common GSM standard has been so successful that it has been adopted worldwide with in excess of 1.2 billion GSM customers now using the service (GSM Association, 2004), most with the capability to use SMS as a standard feature.

Whilst the expansion of GSM internationally gathered pace, the progress with SMS appeared less certain. Indeed coupled with the design of mobile phones, the SMS protocol imposed some serious restrictions and raised glaring flaws in usability (although of course it should be remembered that the protocol was not originally designed to be used on a mobile phone, rather it was expected that a call centre would intercede, as explained later).

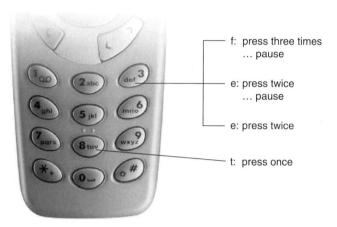

f: press three times
... pause

e: press twice
... pause

e: press twice

t: press once

Figure 4.1. Steps required to input "feet".

For example, the bandwidth limitations of the signalling layer, alongside the constraints imposed by the common standard, set the well-known limit of 160 characters per message; a limit that is reduced to 70 characters for messages written in the non-Latin alphabets, such as Greek or Chinese.

Moreover, with almost all mobile phones being based on the standard 1 to 9, plus 0, keypad layout, the composition of messages was cumbersome and time consuming. On phones with these keypads, several characters are associated with each key, meaning that numerous key presses can sometimes be required to enter one character. To enter "v", for example, the "8" key must be depressed three times. Furthermore, to write two characters from the same key consecutively requires a pause in the procedure. Figure 4.1, for example, shows the steps required to input "feet". First, the "3" key must be pressed three times to enter the "f" character. A short pause is then necessary before the "3" key can be used to input the first "e" (rather than change the already typed "f" character). The pause allows the phone to recognise that a new character is to be inputted. A further pause is then needed before the "3" key can be pressed again to input the second "e". Finally, the "8" key is pressed once to input the "t".

As SMS increased in popularity, various solutions were developed to overcome these limitations. The ability to concatenate messages offered by some operators and available on specific phones provided a way for two or more messages to be strung together and read as one. Aiming to ease message composition, phones with innovative form factors and a larger number of keys were also introduced and predictive text-entry systems were introduced. Predictive text-entry allows characters to be inputted using a single key press and avoid the need for the laborious multi-tap method. Put simply, the system employs a dictionary with which it "predicts" the most likely word that has been entered from consecutive key presses. Leading manufactures such as Nokia and Ericsson (now Sony Ericsson), for example, have produced a succession of devices designed to support single-tap character entry.

4.4 "Texting" in the Making

With its limitations and the apparent primitive character of the technology, it was hardly surprising that the mobile phone operators and manufacturers had no strong business model for SMS. The emphasis for the launch of GSM was on the delivery of talk and international roaming; its unique selling points were the ability to use your own mobile phone anywhere in Europe and improved security and quality of service. The operators' vision for SMS was limited; its broad-based appeal was initially as a unidirectional system for sending "mobile terminated" messages to customers, such as voice mail notifications. Early SMS campaigns to promote the delivery as well as receipt of messages, rare as they were, almost exclusively targeted at business users and positioned the service as a second-rate add-on to voice transmissions. Unlike Wireless Application Protocol (WAP) (that was to be the next wave of data services promising a direct line to Internet content), the decidedly unsexy SMS was of little interest to an industry bent on promoting itself as exclusive and futuristic (Cooper *et al.*, 2000). In this climate, the industry was caught largely off guard by the upsurge in SMS usage (particularly amongst young non-professionals). From the early days of SMS, several significant milestones appeared to have shaped the transformations in business strategies adopted by the industry players in their efforts to capitalise on the unanticipated windfall.

The commercial viability of SMS over GSM owes much to the earlier successes of the pager and related paging services. Although the underlying technologies behind paging and SMS differ significantly, it was the popularity of the paging services and, in particular, the charging models that first struck a chord with businesses in the mobile telephony sector. Having been handed the capacity for SMS, mobile telephony operators saw the potential to offer both voice and data services and, crucially, offer an integrated alternative to paging solutions.

The publicity surrounding SMS, prior to the launch of GSM in 1991, offers an indication of just how this view was presented by industry players. In Geneva, at Telecom 1990 (an event held every 4 years), marketing brochures from the organisation of GSM network operators, known as the GSM MOU – now subsumed within the new global industry body www.gsmworld.com – suggested that SMS was seen as a mirror image of paging.

"This (SMS) service allows the transmission of messages up to 160 alphanumeric characters to be sent to a subscriber. This can be seen as an advanced form of paging, but has a number of advantages. If the phone is switched off, or out of the area covered by GSM, the message is stored and offered to the subscriber when he reappears. This gives a much greater confidence that it has been received. Also, the user needs only one piece of equipment (the mobile phone) and the caller needs to know only one number (the mobile phone number) for telephony and paging. Some phones will even be equipped for originating these messages, but it is expected that generally telephony will be used to call an operator who types in the message at a Service Centre" (GSM MOU Permanent Secretariat, 1990).

There are some clear similarities between the GSM MOU's description of the SMS, written in 1990 (well before the first SMS enabled phones had been launched), and the now ubiquitous texting phenomenon 15 years later. However, what now seems striking and somewhat peculiar is the idea that messages were not considered to be something people would compose themselves on their mobile phones. Taking a cue from the paging model, the organisation's brochure presents a model of messaging in which a service centre's operator would act as the primary vehicle for composing and delivering messages (even though it was recognised phones could be equipped to do so). The very idea that the 16 billion text messages that were sent in 2003 in the UK (Mobile Data Association, 2004) could each have been called into an operator who typed it up and sent it onto each mobile phone seems preposterous. But in 1990 it seemed an obvious solution: one that drew on the earlier success of paging and that overcame the need to manually enter text using the unwieldy multi-tap keypads. In short, SMS was seen as the cellular equivalent of paging services and the intervention of a human interlocutor, a natural development.

Despite the general acknowledgement that messages could originate from mobile phones, as well as via human-operated service centres, mobile phone operators largely underplayed the concept. Predictably, success for the mobile phone operators was seen to hinge on the charging structures and phone-to-phone messaging was thought to be a loss leader in this respect (not least because of the higher infrastructure costs this involved). Whereas paging had been charged for by calls to the service centre, mobile phone operators struggled to determine how money could be made from the delivery of messages; to set up the billing arrangements and to record and charge for each short message sent was considered by some to be more costly than the revenue it would generate, especially when considered alongside the economics of the paging industry.

It was only through the eventual delivery and uptake of SMS that this mindset and penchant for a paging-like service was displaced. Mobile phone operators soon realised that the setting up of service centres staffed with operators to handle the messages would be restrictively expensive: most mobile phone operators were already finding it difficult to keep up with demand and manage calls to their operators on general phone enquiries and they were unwilling to repeat this experience. In this light, a service that required no human intervention between the sender and receiver of the communication presented an attractive proposition. This provided the impetus for the reworking of SMS and at the vanguard of this shift – a move that turned out to be highly lucrative for the operators – was the enabling of a system to charge pre-pay customers for messages.

As SMS was thought unlikely to gain mass appeal, little initial effort had gone into establishing a model for charging those using pre-pay or "pay-as-you-go" options; pre-pay packages from the network operators made no reference to SMS and, more importantly, no mechanism had been put in place to link pre-pay billing with the use of the short-text message service. With some operators, this loophole in infrastructural and charging schemes gave pre-pay users the opportunity to send messages for free. However, with customers and particularly younger users exploiting this gift horse, the operators quickly moved to install the necessary technological and billing systems. Despite their relatively rapid response, this inadvertent oversight by some operators may well have contributed to the early uptake of SMS and also had some bearing on its appeal amongst younger users, the largest consumers of pre-pay packages.

Alongside the move away from the paging model and the investment in an infrastructure to charge pre-pay customers, further factors played a significant contribution in broadening the appeal of the SMS service. Initially, messages could only be sent between compatible mobile phones on the same network. It was not until 1998 that the incumbent UK operators agreed "interconnect" arrangements to allow for messages to be sent regardless of the network operator to which the phone was attached. It was possible to send messages between some international networks before the same compatibility was achieved between network operators in many of the countries offering GSM service.

This transition was smoothed and, indeed, made possible because both the infrastructure and the mobile phone manufacturers had had to comply with the GSM standards that required a common interface between devices and networked systems. Moreover, it was a commonly agreed upon system of charging for international roaming that made it possible to send and receive messages from any regions with GSM coverage, and from any GSM compliant device.

4.5 The Confluence of Actors

So far, this chapter has presented brief accounts of the SMS technology and the role the mobile industry's business interests played in setting the scene for the widespread adoption of texting. The development of the GSM standard and its acceptance first within Europe and then internationally was shown to lay the foundations for SMS. Specifically, the signalling channel freed up on mobile phone networks provided the spare capacity for the text-based messaging service. This technological potential was consolidated with a shift in the industry's operational model from one based on a paging-like arrangement to phone-to-phone flat rate charging and, crucially, the incorporation of pre-pay packages into the billing system. Industry-wide cooperation was shown to play a similarly important role. Both the interconnect agreements between operators and the common standards designed to enable international roaming ensured that SMS did not languish as a marginalised, proprietary service.

These accounts reveal how a mixture of factors has each impinged on the emergence of SMS. There is paging: that is, its past popularity, its billing and pricing schemes, etc; the economics of messaging delivered over GSM for the mobile phone operators; the eventual delivery of SMS and its growing body of users; and the collective decisions and arrangements made by the major bodies and companies in mobile telephony. Of course there are a host of factors not mentioned thus far. For instance, there are the regulatory and policy arrangements acting to moderate and control the progress of GSM and its offspring, SMS, within and between the initial participating countries. There are also the tensions between standardisation and competition, forcing companies to define their own best interests and the best interests of the industry at large. This was exemplified in the UK where slow progress was made in formalising the interconnect agreement for the delivery of text messages between phones on different networks. Not to belabour the point, what is of particular interest here is that each of these factors, acting en masse, writ large and writ small, came to constitute an emerging heterogeneous collective, weaving together in different ways, negotiating their positions and ever so gradually transforming the capacity for the primitive exchange of digitally encoded texts into the short-text messaging service.

What has so far been only alluded to is the role the consumer has played in this collective of heterogeneous actors. Central to the argument in this chapter is the thesis that the above developments were tightly interleaved with people's initial experiences with SMS and the service's subsequent and largely unexpected uptake. One of the first signs of this was the industry's response to the initial increase in use of text messaging once a "point-to-point" delivery and receipt system between phones had been put in place. With the onset of the point-to-point system increasing data traffic by as much as 25% (Netsize, 2003), providers of the service realised that phone users were perfectly capable of composing and sending messages as well as receiving them and were, if anything, drawn to a system that had no human interlocutor. The attraction to this point-to-point arrangement quelled the plans for the paging model that operators had first latched onto (although interestingly a remnant of this remains in the shape of the SMS Centre, an electronic solution for managing and routing messages between phones and networks).

4.6 The SMS Argot and New Social Practices

The need to manually enter text also presented a somewhat unexpected twist to the adoption of SMS. Paradoxically, the limit of 160 characters and the cumbersome and time-consuming multi-tap method for entering text on phones struck a chord with users, particularly younger ones. Abbreviations, acronyms and text-based emoticons (such as CUL8R, LOL and smiley faces, e.g. :-) , ;-) , etc.) adapted from Internet messaging were brought to texting, largely by younger users, in order to ease message composition. To counter the limit on message length, texting-specific conventions were also devised such as the scant use of spaces and the omission of subject pronouns like "I" or "he" (Hård af Segerstad, forthcoming). As well as bringing an informality to SMS communications, this hybrid of styles has established new linguistic repertoires that allow for the intimacy afforded in face-to-face encounters to be reproduced between physically remote interlocutors (Ling, 2003; Thurlow, 2003).

A closely related side effect to the development of these linguistic practices has been the gradual introduction of a new and unique SMS argot (Vincent, 2004). The conventions that first appeared in order to resolve the techno-logical constraints of text-entry became a means of demonstrating a com-petency in texting and one's inclusion in a valorised youth movement surrounding mobile phones (Green, 2003). Testament to the argot are the myriad books now available each explaining and instructing us on the particularities of the texting shorthand, aptly titled "WAN2TLK", "Uwot", "TxtJox" and so on. Further evidence lies in the persistence of the shorthand, despite the introduction of predictive text-entry solutions and message concatenation. Teenagers have been particularly slow to give up their new vernacular (Kasesniemi, 2003) and – flouting the arguments for efficient interfaces – the Digital World Research Centre has found, in our own field studies, occasions in which messages have first been composed using predict-ive text-entry and then reworked to incorporate the texting jargon, once the predictive system has been turned off.

Tightly coupled with the new argot, the point-to-point model not only acted as an initial driver in the SMS service's mass appeal, but also acted as a catalyst to strengthen its position still further. Having inadvertently con-vinced the operators and manufacturers to choose the point-to-point solu-tion through their higher than expected usage rates, SMS users of various persuasions transformed the service into something that could offer more than just practical value.

Mirroring the peer-to-peer services available on the Internet, the point-to-point model came to provide a mobile solution for the asynchron-ous delivery of messages and, crucially, gave users the ability to store both sent and received messages on their own devices (in contrast to voice messages that are stored on operators' remote servers). On a practical level the asyn-chronous messaging service transformed the way people were able to organise such things as appointments. Arrangements could be left to the last minute or coordinated on a moment-by-moment basis to suit delays or changed plans (Ling and Yttri, 2001; Ito, forthcoming). The same principles were even applied to coordinate collective action (Rheingold, 2002), as in the case of the ousting of Philippine President Joseph Estradaor (Bociurkiw, 2001) or during the UK's fuel price protests in September 2000 (MacLeod, 2000).

Limited as it initially was, the storage facility on phones that was needed to support the point-to-point model was fundamental because it provided a basis for the transformation of messages from mere digitally encoded signals to saved mementos of important personal interchanges. The standard header information on SMS messages along with the address book's integration with caller ID lent itself to this as it tagged each message with the date it was sent and the sender's name (if in the address book). Having labelled messages, carefully crafted in the texting argot, that could be stored on the phone, ensured that, for some at least, particular messages could be kept and treasured as the embodied residue of relationships or, equally, disposed of when relationships soured. The restrictions on storage capacity, particularly on earlier phones, served to enhance this sense of the importance of messages because it meant that only the most valued or cherished could warrant being kept. The mechanics of SMS allows messages to be personally and privately inscribed, delivered at will and responded to in kind, along with the capacity to store the more important exchanges. This goes some way to explaining what for some has become the intimate and hugely sentimental character of texting. The practices of reciprocity and the embodied give and take in messaging are closely aligned with the anthropological depictions of gifting that has, at its heart, the object of sustaining intimate social relations (Taylor and Harper, 2003).

The billing system has not been immune to this complex interplay between the social, economic and technical aspects of SMS. On the face of it, the flat rate and relatively small charge for sending a message have been explained as reasons enough to opt for the sending of an SMS over the making of a voice call (when the pleasantries of talk are considered unnecessary). In the UK the cost to the user of sending an SMS has remained more or less stable at a nominal 10 pence; set alongside a tariff for voice calls that was originally three times that figure the motivations and drivers have appeared simple enough. Even teenagers who have been continuing their unabashed usage of the service claim the management and reduction of cost to be one significant reason for its popularity (Grinter and Eldridge, 2003). On closer inspection, however, the issues surrounding charges appear less obvious. Taking into account the message's essence as a treasured item, or gift, it would seem that its monetary value has some bearing on a sender's social standing and thus cannot be simply glossed over through naive economic explanations. Teenagers' measures of frugality stand as testament to this; messages sent over the Internet using free services are deemed cheap amongst teens and, worse yet, those who are thought to be too mean to reply to a message at all are subject to all out exclusion (Taylor and Harper, 2003). Such social pressures can mean protracted conversations over SMS between teenagers, thereby incurring bills they appear unwilling or unable to curtail.

The importance of a message's social value has not, of course, meant that texters (as they are colloquially known) have settled for paying for their obsession. Users, and particularly teenagers, have shown they are aware of the complex billing offers operators have used to lure its customers and throw off the competition (Grinter and Eldridge, 2003). This was evidenced during the fierce price wars and offers of "free" minutes and "free" texts that took place in the UK in the late 1990s. These business adjustments gave the more astute SMS adoptees – again more often than not those youngsters from 16 to 25 – a means of maximising free message quotas by juggling what service providers they used and when they used them. Rather than carry multiple phones, texters could switch between their all-important SIM cards (Subscriber Identity Module), a practice that played its part in the spawning of a black market for the exchange of phones in playgrounds, school canteens and other less salubrious environs.

4.7 Whatever Next?

The outcome, so far, of the unfolding SMS history has been the transformation of a product from a technologically designed concept to a mass-market service generating unprecedented revenues for the industry. According to Mobile Data Association (2004) it represents 16% of total revenue for the UK mobile telecoms industry and amounts to somewhere in the region of 16 billion messages sent each month in 2004. Over its 7-year lifespan, SMS has been a boon for the mobile phone industry and it has carved out a weighty place in the hearts and minds of consumers.

This chapter provides only a tiny glimpse of all that has occurred to make SMS what it is today and much of it in the last section is skewed towards research into young people (incidentally, where the majority of social science research has thus far been undertaken). The broad point to be gleaned is that the SMS history is full of remarkable twists and turns. At no point is it possible to pause and hope that a single snapshot can account for all that has come before it or, much as the industry would like, what might follow. Indeed, the relations between the assemblies – the imbroglio of human actors and their non-human counterparts – continue to grow increasingly complicated over time. The history continually unfolds from multiple standpoints where trajectories are set through unpredictable and sometimes ungainly associations.

With this multi-threaded history of SMS, more often than not the industry has had no choice but to respond to the market with alacrity and has been given little opportunity to drive, let alone gain insight into, how the broader mobile messaging market might take shape. Any implied scepticism for complete and accurate renditions of the unfolding history of SMS should not, however, be taken as a surrendering to some inexorable fate. Matters of some importance can be taken from the broader point of the presented SMS history that might offer the mobile telephony industry a juncture from which to reflect and answer "whatever next?".

The SMS history offers important lessons for those trying to make the most of what can be best described as the "third-generation (3G) challenge". The 3G mobile communication companies have landed themselves in all sorts of trouble with the massive fees they have paid out for network licenses – fees they may find hard to recoup – the technological difficulties they must confront in implementing a robust network and very little idea of what 3G services might be attractive to customers. The surprise successes surrounding SMS reveal that technological innovation, market forecasts and consumer understanding are far from sufficient in predicting how to produce profitable solutions. With hindsight in its favour, the SMS history reiterates what many a good business model has incorporated; it offers a further reminder of the complex interplay that exists between all actors operating within a market. More importantly, perhaps, it suggests that, by intermingling with people, things, like text messages, mobile phones, cellular networks and business partnerships are transformed into so very much more.

This perspective might help to give us some insight into the adoption of two interrelated offerings that both manufacturers and mobile phone network operators hope will enable them to make good on the 3G venture, namely camera phones and picture messaging. At this stage, it is unclear why, but with the launch and big push on the promotion of camera phones in 2003/2004, the suppliers of picture messaging did not ensure the compatibility of devices and interoperability between networks. Although compatibility and interoperability have improved, the service continues to be troublesome for users. These problems have been compounded because the operations needed to send picture messages are not at all clear and the billing arrangements can be confusing with charges based on the size of pictures sent. Whatever the reasons for these problems, the industry's players probably hope that picture messaging will gain in popularity once the technological issues and interconnect agreements have been ironed out (as happened with SMS).

There is one key point that may put sway to this, however, that hinges on the complex interplay of mobile telephony and its uptake amongst users. Unlike SMS, where it was the consumers who played a strong part in driving technological and business developments, the industry has been on the offensive with picture messaging and it has promoted the service through a number of large campaigns. In light of these efforts, consumers' lacklustre interest in picture messaging suggests that the service is an altogether different beast than SMS. What the industry appears to have assumed is a simplistic determinism, relying on the tried and tested "if-we-build-it-they-will-buy-it" model of technological progress. The trouble is this standpoint on technological development ignores the real-world complexities that led to the proliferation of SMS; it glosses over the fact that what contributed to the popularity of SMS was the tight coupling of technological and business developments and their intermingling with people's practical usage of the service and its unique properties – its 160 character limit, textual content, etc. and its complete lack of sophistication.

Upon closer inspection, with particular attention given to the relations between camera phones and people's everyday practices, there are a number of reasons why the adoption of picture messaging cannot be seen as equivalent to the uptake of SMS. For instance, although the sales of camera phones have been high and early research indicates picture-taking with phones is popular there remains a relatively low proportion of pictures actually being sent; the simple showing of pictures to those nearby is by far the most common use of the picturing features on phones (Daisuke and Ito, 2003; Lasen, 2004). Given that camera phone users have had difficulty finding others with compatible devices or messaging that works, this should hardly be surprising. What might be less obvious is that this practice of show-and-tell is not only something that predated picture messaging, with co-proximate phone users frequently sharing and comparing their text messages (Weilenmann and Larsson, 2001; Taylor, forthcoming), but is also something deeply ingrained in how we have learnt to orient ourselves around pictures, whether they may be on paper or rendered digitally (Chalfen, 1987; Frohlich *et al.*, 2002).

A further point is that it would appear that the pictures taken using camera phones have a quality that distinguishes them markedly from SMS messages (and photographs more generally). The camera phone pictures tend to have a mundane character to them (Daisuke and Ito, 2003). Typically, they will be of everyday objects or events, such as a meal, pet, the work or school commute, etc. This suggests that the pictures taken on phones end up serving a very different purpose to the usual content of a text message. The latter, as we have seen, tends to be ascribed importance through the textual content and the reciprocal back and forth that can ensue. Thus the message can come, through its very form, to embody intimacy and social ties. Picture messages, on the other hand, appear to be centred on capturing and representing a person's prosaic movements through time and space; they preserve the short lived, momentary glimpses that make the banal extraordinary (Koskinen, 2004), with the phone's memory providing a means of making them immediately retrievable. The phone thus acts as a repository allowing intimacy and social connection to be achieved later, through opportunistic show-and-tells. Unlike SMS, what picture messaging tends not to be used for is to send carefully composed content (i.e. pictures) of one's self or one's participation in a special event. From a Westerner's perspective, at least, there seems something distasteful in such showmanship: a narcissism in the unwarranted "look-at-me!" quality of a message that is delivered to a recipient. As Koskinen (2004) has shown in his study of Multi-Media Messaging (MMS) use, this "posturing" and one-upmanship can be used in picture messaging as a means to attract attention and encourage a response. It seems an unlikely and probably unpopular method to sustain one's social standing, however.

Of course, picture messaging will no doubt be used to send point-to-point messages, much as SMS has, and there are reported trends in this direction (Koskinen, 2004; Ling *et al.*, forthcoming). What is relevant here and arguably more interesting is that the social practice of face-to-face picture sharing that has emerged has a good deal to do with the design of the mobile phone and the implementation of the picture messaging service. The phone's size, portability, "at-handedness" and plain ubiquity all work together to render it a highly personal device that can be used to take snap shots at will, store the memorabilia of a day's routine and show pictures to others who are close by. The idiosyncrasies of the service's point-to-point model (unchanged from SMS) mean that pictures are sent from one personal device to another so that there always appears to be the brashness of "look-at-me!", no matter how unintended. The billing model for the service also plays its part, of course. Experience of SMS has demonstrated that it is not so much the total cost of messages sent that dictates the use of the service, but rather how visible the cost is; because of their fixed price, text messages have a clear price tag. Picture messaging on the other hand – with its close association with 3G billing systems that are based on the amount of data sent and received – offers a far more complicated and, consequently, less appealing proposition to consumers.

These issues indicate picture messaging could benefit from quite a different service model. The trouble operators and manufacturers face is how to invigorate the practice of sending picture messages through a service that either augments the current ways pictures are shared or produces some entirely new association with pictures. For example, a service can be envisaged that augments show-and-tell so that a picture's significance becomes something worked up between two or more parties rather than simply and unceremoniously delivered to the phone's inbox. The features that are distinct to picture messaging because of the particularities of the camera phone and service must be oriented so that people feel that they are participating in the common experience of showing and talking up pictures.

A solution that accommodates show-and-tell, then, should do more than merely aim to support the collective practices of picture sharing by, for example, enabling text (or other media) to be combined with sent pictures. It should also aim to re-configure the point-to-point system. For instance, as Frohlich *et al.* (2002) have suggested in their more general studies of photography and "photo-talk", pictures take on meaning through the reminiscing and story telling that accompanies shared viewings. A solution might thus be to provide a central repository for pictures where they can be viewed using mobile phones and commented on, much like the web-logs (colloquially known as "blogs") that have become an increasingly popular method for presenting and discussing personal and often mundane accounts of daily life on the Internet. Naturally, the interface issues on small-device displays would have to be thought about carefully for such a service, but the general point is that this solution seeks to move away from the point-to-point model that appears to be unsuitable for at least some forms of picture sharing.

Again, what remains of overall importance are the tightly knit and unfolding relations between any intended service and its use. One single factor cannot be seen to determine another. Rather the relations between a network of interacting agents and how they lend themselves to promoting one historical trajectory over others has to be taken into consideration. For the suggestion briefly outlined above, the device, service and social arrangements have been thought through in relation to one another and set out to promote the common practice of show-and-tell.

4.8 Conclusion

The impetus for this chapter has been the desire to demonstrate how the relations between multiple entities intertwine in the ongoing development of technologies such as SMS. Although much of the material drawn on exists in the mobile phone-related literature, our intention has been to bring a number of seemingly disparate sources together, in order to reveal how the multiple bodies, agencies, actors, things, etc. in a market operate in a collective process of production. Thus, what we have aimed to convey are the ways in which a heterogeneous collection of entities become entangled to make ever more complex networks and transform technologies into a great deal more than lifeless objects.

The SMS history has provided a useful device for this objective because it has amassed so many entities and interrelations between them over its short lifespan. The unfolding relations have been, from the outset, crucial in shaping the messaging service's history. Industry-wide agreements were interlinked with business models to form the basis for worldwide interconnectivity. These same agreements provided the scope for what was thought to be an inconsequential data channel over which short-text messages could be at first received by mobile phones and later, through infrastructural changes, composed and sent. The mobile phone operators, along with their business models, fell back on the earlier successes of paging to envisage a service staffed by human operators targeted at business customers. Meanwhile, a grassroots user base emerged made up, largely, of young non-professionals. These young users were able to sustain mobile phone ownership because of the pre-pay options the phone operators had made available and the flat rate and relatively low charge for sending SMS messages provided a popular means of managing costs. The technological constraints of mobile phones and the service, itself, gave rise to a texting shorthand amongst the more regular SMS users and gradually this evolved into a new argot, binding a collective of texters together through their common modes of expression.

Of course, the story continues and much more has been bound into the collective than presented here. What we catch sight of, though, is how it is that essays get written in hieroglyphics and mobile phones can become the cause of falling standards in literacy. We see the emergence of things that have agency, of technologies having their say through their inexorable part in the collective. It is this lesson that has been used to contemplate a history for picture messaging and camera phones. We have sought to dismantle the divide between the technological and social, the non-human and human, and reassert the collective by placing the picture messaging service in and amongst what people do. While this small example cannot do justice to all that is in store for picture messaging, what it serves to illustrate is that we have come too far with both SMS and the ways in which we handle pictures to hope for the repetition of the SMS history. A new history is upon us.

References

Bociurkiw M (2001) *Revolution by Cell Phone*. Forbes.

Callon M, Rabeharisoa V (2003) Research "in the wild" and the shaping of new social identities. *Technology in Society*, **25**: 193–204.

Chalfen R (1987) *Snapshot Versions of Life*. Bowling Green State University Popular Press, Bowling Green, Ohio.

Cooper G, Green N, Moore K (2000) *Mobile Culture: The Symbolic Meanings of a Technical Artefact. British Psychological Society Conference*. London, 20 December.

The Daily Telegraph (2003) Girl writes English essay in phone text shorthand. [3 March]. http://www.telegraph.co.uk/news/main.jhtml?xml=%2Fnews%2F2003%2F03%2F03%2Fntext03.xml

Daisuke O, Ito M (2003) Camera phones changing the definition of picture-worthy. *Japan Media Review*, http://www.ojr.org/japan/wireless/1062208524.php

de Laet M, Mol A (2000) The Zimbabwe bush pump: mechanics of a fluid technology. *Social Studies of Science*, **30**: 226–263.

Frohlich D, Kuchinsky A, Pering C, Don A, Ariss S (2002) Requirements for photoware. *Conference on Computer-Supported Cooperative Work*, CSCW '02. New Orleans, Louisiana, ACM Press; pp. 166–175.

Green N (2003) Outwardly mobile: young people and mobile technologies. In: Katz JE (ed.). *Machines That Become Us: The Social Context of Personal Communication Technology*. Transaction Publishers, New Brunswick, NJ.

Grinter RE, Eldridge MA (2003) Wan2tlk? Everyday text messaging. *Conference on Human Factors and Computing Systems*, CHI 2003. Fort Lauderdale, FL, ACM Press, pp. 441–448.

GSM MOU Permanent Secretariat (1990) *A Guide to Pan-European Digital Cellular Radio MOU-MP*. Document 4, Version 3.1.0, October.

GSM Association (2004) http://www.gsmworld.com, August.

Ito M (forthcoming) Mobile phones, Japanese youth, and the re-placement of social contact. In: Ling R, Pedersen P (eds). *Mobile Communications: Re-negotiation of the Social Sphere*. Springer-Verlag, London.

Kasesniemi EL (2003) *Mobile Messages: Young People and a New Communication Culture*. Tampere University Press, Tampere.

Katz JE, Aakhus MA (eds) (2001) *Perpetual Contact: Mobile Communication, Private Talk, Public Performance*. Cambridge University Press, Cambridge.

Koskinen I (2004) *Seeing with Mobile Images: Towards Perpetual Visual Contact. Conference on The Global and the Local in Mobile Communication: Places, Images, People, Connections*. Budapest, Hungary.

Lasen A (2004) *Affective Mobile Phones. Fifth Wireless World Conference: Managing Wireless Communications*. Guildford, Surrey.

Latour B (1993) *We Have Never Been Modern*. Harvester Wheatsheaf, London.

Ling R (2003) The socio-linguistics of SMS: an analysis of SMS use by a random sample of Norwegians. *Front Stage–Back Stage. Mobile Communication and the Renegotiation of the Social Sphere*. Grimstad, Norway.

Ling R, Julsrud T, Yttri B (forthcoming) Nascent communication genres within SMS and MMS. In: Harper R, Taylor AS, Palen L (eds). *Inside Texting: Social and Design Perspectives on SMS*. Kluwer Academic Press, Amsterdam.

Ling R, Yttri B (2001) Hyper-coordination via mobile phones in Norway. In: Katz JE, Aakhus MA (eds). *Perpetual Contact: Mobile Communication, Private Talk, Public Performance*. Cambridge University Press, Cambridge; pp. 139–169.

MacLeod A (2000) Call to picket finds new ring in Britain's fuel crisis. *Christian Science Monitor*, http://csmonitor.com/cgi-bin/durableRedirect.pl?/durable/2000/09/19/p7s2.htm

Mobile Data Association (2004) http://www.mda-mobiledata.org

Netsize (2003) *European SMS Guide: Enabling Mobile Business and Entertainment*. Paris, France.

Rheingold H (2002) *Smart Mobs: The Next Social Revolution*. Perseus Pub., Cambridge, MA.

Taylor AS, Harper R (2003) The gift of the gab? A design oriented sociology of young people's use of mobiles. *Journal of Computer Supported Cooperative Work (CSCW)*, **12**: 267–296.

Taylor AS (forthcoming) Phone talk. In: Ling R, Pedersen P (eds). *Mobile Communications: Re-negotiation of the Social Sphere*. Springer-Verlag, London.

Thurlow C (2003) Generation Txt? Exposing the sociolinguistics of young people's text messaging. *Discourse Analysis Online*, **1**.

Vincent J (2004) The social shaping of the mobile communications repertoire. *Journal of Communications Network*, **3**, Part 1, Jan–Mar.

Weilenmann A, Larsson C (2001) Local use and sharing of mobile phones. In: Brown B, Green N, Harper R (eds). *Wireless World: Social and Interactional Aspects of the Mobile Age*. Springer-Verlag, London and Heidelberg, pp. 99–115.

Part 2
Present Uses

5

Emotional Attachment to Mobile Phones: An Extraordinary Relationship

Jane Vincent

5.1 Introduction

In this chapter it will be argued that there is an extraordinary relationship between people and their mobile phones; one that embodies the emotional attachment many people have to their mobile phone and all that it engenders. It will be asserted that these emotional behaviours arise out of the omnipresence of mobile phones (mobiles) in society today and the synergy between people's behaviours and the capabilities of their mobiles. This chapter describes these behaviours, both in the way that people are observed using their mobiles and in the terms they use to describe their relationship with them. It offers evidence, from empirical research conducted by Vincent and Harper (2002) and Vincent and Haddon (2003), that shows just how attached people have become to their mobiles and it offers some explanations for these behaviours. Finally, this chapter suggests some of the implications that these behaviours might have on the development of mobile communications and how these are different from that of any other information communication technology devices.

5.2 The Omnipresence and Absent Presence of Mobile Phones

Although mobiles have been in everyday use for less than 20 years they are already omnipresent in many countries with increasing ownership and usage that is growing on a daily basis. It is reported on the GSM Association web site (http://www.gsmworld.com) that in 2004, there were in excess of 1.2 billion mobile phone users worldwide. Omnipresence of mobile phones occurs in societies when everyone becomes aware of them even though they may not own or use one themselves. This omnipresence of mobiles has been influenced by the behaviours of early adopters who themselves would have drawn on their experiences from other communications devices. Although a public mobile phone service has been generally available since the 1980s, usage by the majority of the population in countries such as the UK has occurred only since the late 1990s. Thus society's experience of mobile phone communications is relatively recent. Although new mobile communications systems replaced antiquated mobile telephony services overall the mobile phone service was an addition to people's lives. People had not previously been able to make voice calls on the move, in locations where there was no fixed phone or have a number personal to them; with a mobile phone they can now do all of these things and more: such as texting, accessing the Internet, taking images and video, and playing music and games.

The mobile phone was so different from any communications device that came before that initially, there was no established etiquette for its use. Some suppliers of service have provided a guide to mobile etiquette such as BT Cellnet in the UK (1994) and *Siemens Guide for Business Users in the UK* (2004). However, more usually, people would refer to their own experiences of private mobile communications services (such as CB radio or ship-to-shore) and fixed telephony services, as well as to their peer groups, to establish a modus operandi. As with the introduction of fixed line phones, there were numerous myths about the service. As described in Chapter 2, people thought that fixed phones would cause a variety of problems, such as aural overpressure and even lead to hysteria; these anxieties are mirrored in concerns about mobile phone health risks, such as cancer. Initially used only by the cash rich and businessmen, mobiles are now accessible to most people. In his work on the "domestication" of the service Haddon (2003) refers to the situation where now "one need not clarify why one has a mobile telephone; rather one is called to answer if they do not". To achieve this turn from it being a novelty to have a mobile to it being a novelty not to have a mobile requires a new set of normative behaviours common to all users. Goffman (1956) discusses developing these normative behaviours, in his work *The Presentation of Self in Everyday Life* in which he articulates the actor's ability to adapt to something new by creating a "front". The front of a mobile phone user continues to develop, shape, and be shaped by the technologies that provide the variety of mobile phone services.

In the UK at the beginning of the 21st century, many people respond to mobile phones in their day-to-day lives as if they had never known anything different; consider for a moment who is calling whom? In research conducted by Vincent and Harper (2003) and Vincent and Haddon (2004), it was found that mobiles are used less by business people to communicate with clients and more by business colleagues, family and friends to maintain social relations. Evidence indicated the emphasis is on talking and texting between family and friends to keep in touch. Even in a maturing market people were not yet ready to admit they used the phone just to chat probably because if they did they would have to justify spending money on idle chatter. Perhaps the most obvious use had to do with setting up social and family arrangements: "I call my friends … stupid calls … I'm meeting them in half an hour and I'll call them, speak to them … until I meet them", or "What time will you be home? What do you want for dinner?" or "I'm going to be late – that kind of thing". Another had to do with avoiding making set appointment times by arranging to call when arriving such as when "meeting in a park full of people". Another example was about making or breaking relationships: "You can be silly on texting, you're too embarrassed to phone" and "If I want to speak to my girlfriend any time of the day I know that I can and it kind of takes the fun out of it when I'm seeing her." However, for some the frisson of making or receiving an intimate call in a public place is perhaps tempered by the response of others to the intrusion of that moment. Schegloff (2002) describes a scenario in which a person audibly conducting an intimate conversation over the mobile on a train, calls out in enraged protest "Do you mind?! This is a private conversation!" when the person's gaze is intersected by another. The desire for a personal space juxtaposed with the invasion of the public place with private behaviours has meant that new etiquettes have developed whereby in many public places anything goes but in others using mobiles is not acceptable. Train carriages are provided in which the use of mobiles (and laptops too) are banned and people are asked to switch off mobiles in an increasing number of locations, such as in clubs and at work. Research conducted in the UK by DWRC and Germany Erfut University (Vincent and Harper, 2003) showed that although they do not like it when others behave in this way people will still use their mobiles when driving, in the cinema, theatre, classroom and even places of worship. Cultural differences are emerging regarding these mobile phone etiquettes but the differences are about how far people will go in their use of the mobiles and what is considered polite or impolite mobile phone usage rather than whether they should be using their mobile phone at all. Katz (2003) describes this use of mobile phones in public space as a "choreography" of mobile communications and asks "is the irritation and displeasure that results from the public use of mobile phones comparable to ethics, politics or fashion, all of which can change rather quickly, or to biology, which changes but little over many generations". As will be discussed, the impact of the mobile phone is keenly felt by its owner and by society as a whole. Each mobile user in this choreography has their own set of communicants whose presence is felt and focussed through their attachment in some way to the mobile itself. This might be by text, talk, image, stored or in real time, by being in the phone directory, or by the "missed calls" register and so on. In other words there is a multitude of small worlds intracommunicating rather than one large world intercommunicating. Additionally, whether or not a person is using their mobile or even if they do not have it with them at all it has the role, as Gergen (2002) calls it, of having "absent presence". The mobile, or the thought of the mobile has the effect of making the owner feel that others are with them. That people call people they already know and that their mobiles are filled with the details of people with whom they have exchanged numbers means that each person is part of a community of mutually connected people. It is as if there is a dormant pool of common and linked minds residing in the ether that the press of a button on the mobile can energise at anytime. The absent presence of each member of the community of users on each mobile phone has the effect of keeping people in touch with each other regardless of their actual distance apart or where they are.

What has been observed thus far is that mobile phone usage is mostly about social connectivity: it is people calling people they know rather than people making new contacts; it is people talking, texting, and sending or showing images to each other; it is people using their mobiles to conduct intimate communications in the most public of places; and, as it will be argued, it is the emotional content of these communications that is the driver behind the extraordinary relationship many people have with their mobiles. It is people communicating on a public and a private stage with others from whom they are absent even if only for a fleeting moment.

5.3 Examining People's Emotional Attachment to Their Mobile

If you ask people to talk about their mobile phone and what it means to them they tend to use emotional terminology to describe their views. Table 5.1 shows a summary of concerns about emotion and mobile phones identified by Vincent and Harper in their study (2003) investigating the social shaping of mobile phone communications.

These emotions would appear to be peculiar to the mobile phone insofar as they are expressions of behaviours that have as their focal point the synthesis of mobile and user; the emotion is the outcome of the effect of this combination. For example, the feeling of panic is manifested in whether or not one is absent from the device, a concern that appears to have superseded absence from each other because the presence of the mobile is substituting

Table 5.1 Summary of concerns about emotion and mobile phones

Emotion	Explanation
Panic	Absence from the device; being separated from it
Strangeness	Between those who do and those who do not have mobile phones
"Being Cool"	Chilled out, tuned in to the mobile phone culture
Irrational Behaviour	Cannot control heart over mind, e.g., driving and talking
Thrill	Novelty, multi-tasking, intimacy of the text received in public
Anxiety	Fear and desire: e.g., not knowing and wanting to know about others versus too much knowledge

Source: Vincent and Harper (2003).

for the presence of others. Parting from a loved one, for example, is made easier by the fact that if both have mobiles it is as if they are still together. There is a kind of emotional tethering brought about by owning a mobile; an ever-present invisible line that joins the user of a mobile with their nearest and dearest, or with points of contact, positive or negative. Thus although communications between individuals and their social networks would likely continue without mobile communications the "absent presence" of the mobile phone capability affords increased networking, deeper relationships or simply increased contact. The corresponding loss of that capability, however, engenders strong feelings such that people say they panic if they find they do not have their mobile with them and especially if they think they might have lost it.

The idea that people might choose not to have a mobile is a concept that mobile phone owners accept albeit with an air of disbelief and a feeling of strangeness about the person who has refused to have a mobile. After all, to have a mobile is part of "being cool", being tuned into society and one's peer group. The downsides of mobile phone ownership are the inability to control one's usage such as when in a car "Even though I know I shouldn't and I drive over those mini roundabouts in 4th gear" (Vincent and Harper, 2003), and the anxiety of feeling the need to find out where people are just because they can be called up and asked. However, what do you do with the knowledge gained in this way if it is not what is expected? This is a dilemma facing those developing new services that increase the connectivity between users such as by providing tracking and location information, and is an area that needs further study.

Using the emotional terms discussed here to describe a mobile indicates that for the owner the emotional attachment is not with the device itself but with content and contact it enables and the information stored on it. It is this that is extraordinary about the mobile and that stands it apart from other information communications technologies.

5.4 How Attached are People to Their Mobiles?

The way some people use their mobiles and the way they speak about their mobiles has been described. In both instances it is emotion and emotional terms that are used to articulate the behaviours, but it is the need for social connectivity that creates the vehicle for the emotional content in the first place. In considering just how dependent people are on their mobiles one can start to unpack the layers of emotion to find that it is not just how people talk about mobile phones but how mobiles have become embedded in our daily lives to such an extent that for many people they cannot imagine a life without one. One can surmise from this that people have a distinct and essentially emotional relationship with their mobile phone and all that it engenders. This reflects what the mobile enables them to do in terms of being in touch with those they are close to, in the way that the mobile enables emotional and spontaneous behaviours, and in the ways in which people account for and think about their mobiles. They have a physical response to the device which is usually close to hand and carried about the person, and for some it is the way the mobile is held and touched that affirms their belief that their relationship with mobiles is indeed different from other information and communication technology (ICT) devices. Fortunati (2002), in her paper on the use of mobiles in Italy, discusses the different manifestations of the role of the mobile stating that "it is developing an identity that is much more articulate and complex than that of other communicative technologies in as much as it will absorb, but not replace, other technologies such as Personal Digital Assistants (PDA), personal stereo and laptops". This constant synthesis of technologies coupled with the content delivered by and contained on the mobile phone strengthens the dependency of the owner but increases the tension between needing the mobile and concern at losing it and all that it contains. This is not merely a matter of ensuring lists of phone numbers are backed up and personal messages copied somehow, but also the potential loss of the relationships that the mobile delivers either directly or via their "absent presence" (Gergen) as discussed earlier.

The manifestation of these behaviours is the creation of a value paradox in that people have found that they cannot live without their mobile phone but ironically this sometimes means they value it so much that they do not take it with them for fear of losing it. Thus, though it can be argued that the use of the mobile phone strengthens emotional attachments in so doing it may make the mobile too precious to lose.

5.5 Explaining the Relationship with the Mobile Phone

There are various research papers to do with personal relationships and emotional tethering that give expression to the emotional experiences people are having with their mobiles. Some authors have suggested that it is the invigoration of social and emotional bonds over and above anything else that makes mobile communications uniquely appealing. Our previous research on this subject (Vincent, 2003) cites research in Finland by both Puro (2002) and Kopomaa (2000) who argued that mobile communications have brought Finns together in ways that other technologies have not done. Indeed, they go further and argue that Finnish culture has shifted to be more connected than before. Two more articles that are particularly illustrative are Ling and Ytrri (2002) on hyper-coordination amongst Norwegian teenagers and Taylor and Harper (2003) on gift giving and text in the UK.

The article by Ling and Yttri explains how Norwegian teenagers use mobiles to express and display their identity and it finds that the mobiles enable various techniques for sustaining that identity such as the way they use text as gifts and through their choice of responding or not responding to calls from friends. Taylor and Harper show that mobile phones provide a medium through which young people sustain and augment their social networks thereby tethering themselves to their friends and family. They suggest the practice of using text messages can be thought of as forms of gift giving. Gifts include such things as good night messages, jokes and gossip, and are often saved as mementos by the recipients.

In considering just how dependent people are on mobiles, it begins to become evident that in this relationship between the owner and their mobile there is a constant iteration of emotional value creation. Communications made and received, information stored and ring tones used, all contribute to building a unique device highly personal to the user. There is clearly a need, or a number of needs, being satisfied in the process but is there an explanation for the relationship people develop with their mobile phone device? The value paradox and the basic need for social connectivity have been discussed but this does not address why, for some people, there is such an intensity of the emotional attachment to the mobile phone. This emotional dependence on the device suggests that there is some form of synthesis between the user and their mobile such that neither can function without the other. The combination of mobile phones and the wit of the individuals using them have created a synergy of the technology and human behaviours that allows people to flex and vary what they do. This relationship between technology and human behaviour is difficult to define: is it the human behaviour that is manifesting in the design and use of the technology or is the reverse occurring? The answer is probably that both are impacting each other and causing a gradual change in behaviours that allows for the next generation of technology to be developed. The actual process of this synthesis or cooperation between behaviours and technologies has some interesting parallels with the work of Maturana *et al.* (1974) on self-organising systems and more particularly, their work on what they termed "autopoiesis". Every individual has a personal state of autopoiesis, whether or not they have a mobile phone, and the more changes that occur in their lives, the more unique their personal autopoietic state becomes. Maturana and colleagues' research began in the 1970s, well before mobile phones were in use, and arises from the hypothesis that people are unique self-organising systems each affected by the continuing changes that happen to them. Since the advent of mobile phones some of these changes are a consequence of people's own and other's use of mobile phones. Thus the autopoiesis of mobile phone owners is changed constantly by their ability to reciprocally accommodate and compensate their perpetual use of the mobile phone. This is one facet of their constantly changing life and thus their autopoiesis, but nevertheless one that has had a notable impact on them and on society in general. People's autopoiesis is also affected by the omnipresence of mobile phones; even if you do not have a mobile yourself someone you know will have one and may have your phone number stored on it, or you may call them on it. In explaining the relationship with the mobile phone it would appear that emotional attachment owes as much the development of the individual and their autopoiesis as to the connectivity that the mobile affords. These aspects, combined with the omnipresence of mobile phones, are key to explaining the reasons for people's relationship with their mobile phone.

5.6 Will People Be as Emotionally Attached to Their Mobile Phones in the Future?

Using emotion to maximise the potential of products and services is challenging not least because it demands a level of understanding of customers, existing and potential, that is not normally known. More particularly, it demands knowledge of the purposes of the user's communications and how those purposes deliver the emotional value that is so important to them. As future products and services appear they will most likely be adopted if they satisfy needs that are based on these emotional ties, and achieve some augmentation of person-to-person connectivity albeit with some person-to-information connectivity.

The use of SIM card (Subscriber Identity Module) readers by some Service Providers to collect and store information held on mobiles, phone numbers in particular, is a present-day example of a service that responds to users' fear of losing their mobiles. But it goes only a small way to addressing how to do so, in a way that reflects the emotional value that some users place on information stored on their phone. Much more needs to be understood about how emotional values are delivered so that providers of service can identify ways of leveraging opportunities related to emotion. Failure to understand these emotional values can lead to problems. New services could be jeopardised if they replace or impinge upon services to which the user ascribes an emotional attachment. For example, the threat of losing text messaging and having it replaced by new technologically better services may create considerable resistance. For not only is texting now a key tool in sustaining emotional lives, but storing personal text messages is now a highly valuable element in people's emotional arsenal. Texting may be thought of as simply a communicative technique from the supplier's point of view, but to the user it has values over and above this. The adoption of new form factors may also affect these emotional values. For example, the current size of mobile devices supports constant carrying around. Fortunati (2002) talks of the possibility of mobiles of the future tending to reside near, or even inside the human body for long periods, and of today's small mobiles failing to satisfy needs for larger screens with better resolution. Whatever the outcome, emotion is firmly embedded in mobile phone development and is likely to continue to influence new ideas.

5.7 Conclusions

The omnipresence of mobile phones in society enables constant connectivity between groups of users most of whom were already socially connected. Although the mobile phone has not replaced what people do it has made life easier for them, the dependence that people now have on their mobile phones for maintaining this connectivity is such that some have come to depend on their mobiles and at times too much. People express their use of mobile phones in emotional terms: these terms do not describe an emotional attachment with the device itself but with the content it enables, the relationships it maintains and the information stored on it. The use of mobile phones strengthens and deepens this emotional attachment but in so doing may make the mobile phone too precious to lose, thereby creating a value paradox for the owner.

The mobile phone would appear to be the preferred personal communications device for the foreseeable future although it will most likely develop as a hybrid of other information communication technologies rather than replace them. In so doing it will maintain its uniqueness as an emotional device. The mobile phone is predominately a means for achieving person-to-person communications and so the development of person-to-information services such as location and tracking products may be resisted unless presented as an augmentation of a person-to-person capability. This extraordinary and emotional relationship that people have developed with their mobile phone means that services that leverage emotional attachment, in other words those that recognise the strength of the absent presence of others, are more likely to be adopted than those ignoring the emotional needs of mobile phone owners.

References

BT Cellnet (1994) *Mobile Manners: A Guide to Mobile Etiquette in the '90's*. Telecom Securicor Cellular Radio, UK.

Fortunati L (2002) Italy: stereotypes, true and false. In: Katz J, Aakhus M (eds). *Perpetual Contact: Mobile Communication, Private Talk, Public Performance*. Cambridge University Press, Cambridge; pp. 42–62.

Gergen (2002) The challenge of absent presence. In: Katz J, Aakhus M (eds). *Perpetual Contact: Mobile Communication, Private Talk, Public Performance*. Cambridge University Press, Cambridge; pp. 227–241.

Goffman E (1956) *The Presentation of Self in Everyday Life*. Doubleday, New York.

Haddon L (2003) Domestication and mobile telephony. In: Katz J (ed.). *Machines That Become Us: The Social Context of Personal Communication Technology*. New Brunswick, NJ: Transaction.

Katz J (2003) A nation of ghosts? Choreography of mobile communications in public spaces. In: Nyiri K (ed.). *Mobile Democracy Essays on Society, Self and Politics*. Passagen Verlag, Vienna; pp. 21–32.

Kopomaa T (2000) *The City in Your Pocket: Birth of the Mobile Information Society*. Gaudeamus, Helskinki.

Ling R, Yttri B (2002) Hyper-Coordination via Mobile Phones in Norway. In: Katz J, Aakhus M (eds). *Perpetual Contact: Mobile Communication, Private Talk, Public Performance*. Cambridge University Press, Cambridge.

Maturana HR, Varela FJ, Uribe R (1974) Autopoiesis: the organisation of living systems, its characterization and a model. In: *Biosystems Amsterdam North Holland*, Volume **5**, pp. 187–196.

Puro J (2002) Finland a Mobile Culture. In: Katz J, Aakhus M (eds). *Perpetual Contact: Mobile Communication, Private Talk, Public Performance*. Cambridge University Press, Cambridge; pp. 19–29.

Schegloff E (2002) Beginnings in the telephone. In: Katz J, Aakhus M (eds). *Perpetual Contact: Mobile Communication, Private Talk, Public Performance*. Cambridge University Press, Cambridge.

Siemens Communications (2004) Communication overload makes office workers SAD. Business Etiquette Research. DWRC and SSMR, University of Surrey. www.siemens.co.uk 04.06.04.

Taylor AS, Harper R (2003) The gift of the gab? A design oriented sociology of young people's use of mobiles. *Journal of Computer Supported Cooperative Work*, **12**(3): 267–296.

Vincent J (2003) Emotion and mobile phones. In: Nyiri K (ed.). *Communications in the 21st Century Mobile Democracy Essays on Society Self and Politics*. Passagen Verlag, Vienna.

Vincent J, Haddon L (2004) Informing suppliers about user behaviours to better prepare them for their 3G/UMTS customers. UMTS Forum Report Number 34, http://www.umts-forum.org

Vincent J, Harper R (2003) The social shaping of UMTS, educating the 3G customer. UMTS Forum Report Number 26, http://www.umts-forum.org

Textmates and Text Circles: Insights into the Social Ecology of SMS Text Messaging

6

Donna J. Reid and Fraser J. M. Reid

6.1 Introduction

This chapter is concerned with the social and psychological impact of mobile phone text messaging, or "texting". The increasing and widespread use of texting is revolutionising communication in today's society. It is estimated that 72.1% of people in Western Europe own a mobile phone (Katz and Aakhus, 2002) and that over 1 billion messages of up to 160 characters are sent each month in the UK alone (AOL mobile, 2002). "Mobile messaging is the modern way to communicate. It's instant, location independent and personal. That's why the new mobile phone generation has started to favour messaging, making it one of the fastest-growing segments of the mobile communications industry" (Nokia, 2002). The growth in the volume of text messaging, particularly among young people (Haig, 2002), is a social phenomenon which needs to be explained, and its impact on human relationships and psychological well-being understood.

Since records began in 1999, the use of text messaging has continued to rise. In 1999, the total number of chargeable person-to-person text messages sent across the four UK Global System Mobile Communication (GSM) networks was around 1 billion. This figure rose to 11 billion in 2001. In 2003, over 20.5 billion texts were sent, which averages at about 55 million messages per day compared to 48 million in September 2002 and 36 million in September 2001. The Mobile Data Association (MDA) forecasts that by the end of 2004 the number will be around 23 billion (MDA, 2004). So the use of Short Message Services (SMS) text messaging seems to be still increasing. This phenomenal growth of the use of this new technology demands explorations into the social and psychological impact that it has on the millions of users that foster it.

A clear but untested assumption is emerging that young people are both the driving force behind and, at the same time, the slaves of a growing text messaging culture (Thurlow, 2003). According to Nokia's worldwide survey of 3,300 people (Nokia, 2001), the core mobile phone market is the under-45 age group. Over 80% of those sampled in this survey reported text messaging as the most used function on their mobiles. Other studies have found that, in Britain at least, nearly 80% of 14–16-year olds own mobiles (NOP, 2001; as cited in Thurlow, 2003), and that it is this teen market that dominates text messaging, with 90% of teenagers claiming to text more than they talk on their phones (Haig, 2002). However, despite a small number of qualitative studies of teenagers' use of text messaging (e.g. Kasesniemi and Rautiainen, 2002; Ling and Yttri, 2002; Puro, 2002; Thurlow, 2003), little is known about the psychological impact of texting on social interaction amongst regular users, and on the long-term consequences of texting on the development and maintenance of these relationships.

This chapter reports preliminary findings of a four-year study into the social and psychological aspects of SMS text messaging and presents some results emerging from a large sample Internet survey of texting designed to explore the relevance of Katelyn McKenna's work on relationships formed and maintained on the Internet to text messaging. By focusing on positive aspects of conventional Internet use, McKenna and colleagues have attempted to "set the record straight ... about the actual social and interpersonal consequences of the Internet" (McKenna and Bargh, 2000: 59). McKenna and Bargh (1998), for example, found that regular use of e-mail and participating in user groups had improved the lives of some users, particularly those who experienced difficulties with face-to-face communication; for example, those who are socially anxious, lonely, or who have marginal identities (Bamrud, 2002; McKenna et al., 2002). McKenna et al. reasoned that key features of the Internet, particularly the attenuation of personal information (such as appearance, stuttering, shyness, etc.) would allow greater freedom of self-expression and nurture personal relationships that might not have the chance to develop face-to-face. In fact, McKenna et al. (2002) found that the lonely and socially anxious were better able to express themselves and develop close friendships on the Internet than in the "real" world. Whilst people with extensive social networks and frequent intimate social contacts also use the Internet for social purposes (Birnie and Horvath, 2002), many turn to it as a relatively safe environment in which to form close and meaningful relationships "to make a reality out of their virtual lives" (McKenna et al., 2002: 30).

It is suggested in this chapter that this applies equally well to the regular, social use of mobile phone texting. For a significant number of users, sending a text message may be more important for building and maintaining social relationships than for co-ordinating practical arrangements (Ling and Yttri, 2002). However, certain features of mobile telephony set texting apart from conventional Internet use, as well as resembling it from the user's point of view. Firstly, the immediacy, mobility and perpetual accessibility afforded by the mobile phone allows near-conversational levels of synchronous texting, so that an exchange of text messages can resemble online chat in terms of turn taking and discourse structure (Kasesniemi and Rautiainen, 2002). On the other hand, the QWERTY-driven text interface presents the user with an asynchronous medium similar to e-mail, allowing time for composition and reflection, and the opportunity to manage the way users construct and present themselves in their messages (Danet, 1995; Chenault, 1998; Ling and Yttri, 2002).

It is suspected that it is the combination of these two features – the sociability of the chat room coupled with the psychological distance of e-mail – that lends texting a special, but paradoxical, appeal to a significant number of users. In Thurlow's (2003) recent study of undergraduate text messages, only about one-third of messages accomplished functional or practical goals; the remainder fulfilled a combination of phatic, friendship maintenance, romantic and social functions associated with highly intimate and relational concerns. Text messaging therefore seems to provide an opportunity for intimate personal contact whilst at the same time offering the detachment necessary to manage self-presentation and involvement.

The Internet survey described in this chapter extends McKenna et al.'s (2002) model to SMS texting by exploring how people who prefer using their mobiles for texting ("Texters") differ from those who prefer using it for voice calls ("Talkers"). We suspect that whilst Talkers value the immediacy and mobility of the mobile phone, it is the Texters who capitalise on the social environment created by texting to form and manage close personal relationships. Presented here are the preliminary findings of the questionnaire, with particular reference to measures of phone usage, patterns of communication, and two key individual differences studied by McKenna, namely social anxiety and loneliness.

6.2 Method

6.2.1 Designing the Online Survey Questionnaire

A multi-page, branching online questionnaire including multiple-choice, scalar and open-field questions was developed to gather information on mobile phone ownership and usage, and aspects of relationship development. Questionnaire items were drawn from a wide range of sources and included the complete Interaction Anxiousness Scale (Leary, 1983) and the abbreviated UCLA Loneliness Scale (Russell, 1996) employed by McKenna et al. (2002), together with a version of Parks and Floyd's (1995) Levels of Development in Online Relationships Scale adapted to refer to text messaging. Three focus groups were then run to evaluate questionnaire items and organisation

(O'Brien, 1993). These groups consisted of staff and students at the University of Plymouth. Several changes and additions were made to the questionnaire based on these groups. Finally, a pilot study was conducted with a group of 48 year-12 and -13 school children to refine the questionnaire's structure and format, and test the method of delivery.

6.2.2 The Main Study

The online survey questionnaire was advertised on the Internet at various sites, including Plymouth and other universities' online research pages, chat forums, newspapers and a range of list servers. A link to the study was also hosted by one of the Internet's largest search engines. In addition, the questionnaire was advertised through a distribution list to all students at the University of Plymouth.

On opening the questionnaire, respondents were presented with an information page outlining the purposes of the study, with a link to a consent form. A further link checked for mobile ownership, and transferred mobile owners to the first of four pages containing a total of 143 questions. Non-owners were routed to a shortened version of the questionnaire containing two pages totalling 58 questions. Responses were uploaded to the server on completion of each page.

The full version contained questions concerning demographic information, mobile phone ownership and other background information on mobile use, text/talk preferences, mobile etiquette, phonebook contacts, language usage in text messaging, message collection, experiences with mobile problems and the use of mobiles in relationship development and maintenance. Individual differences in user orientation were assessed by the 10-question UCLA Loneliness Scale and the 15-question Interaction Anxiousness Scale, both of which have reported Cronbach reliabilities of $\alpha = 0.89$.

To address the topic of self-expression, items were adapted from McKenna's "real self" questionnaire to include questions regarding the extent to which respondents' families "would be surprised if they were to read his or her text messages", whether respondents "felt more comfortable saying things in text messages than face-to-face", and in which medium (text messages, phone calls or face-to-face) respondents "felt better able to express their true feelings". To assess the breadth and depth of personal relationships established or maintained through text messaging, 15 questions from the Levels of Development in Online Relationships Scale were used.

The questionnaire incorporates various response formats, including quantitative Likert scale items, yes/no items, multiple-choice items and qualitative open-field items. Responses uploaded to the structured query language (SQL) server were transferred to a Microsoft Access database, held on the department's web server.

6.3 Results

6.3.1 The Present Sample

The sample reported here completed the online survey over a 14-day period between 5th and 19th March 2003, totalling 837 completions, of which 331 were males and 506 females. Ages ranged from 16 to 37 years, with a mean of 23.6 years. Of the total, over 90% (756) were British residents, and over half (478) were students and/or staff at the University of Plymouth.

Of those responding to questions concerning relationship status (827), over half (478) were single, over one-third were in a long-term relationship (220) or living with their partners (71), and fewer than one in ten were married (58). Of those responding to questions about the size of their families, 78 (9%) were only children, 409 (49%), 201 (24%) and 76 (9%) had one, two or three siblings, respectively, whilst (8%) had four or more siblings. Over 10% of the sample (91) had children of their own.

In terms of Internet use, nearly all (97%) reported using e-mail "often" or "very often". However, use of Instant Messaging (IM) was much less widespread: only 366 (44%) reported using IM "often" or "very often". However, nearly all (800, 96%) owned a mobile phone.

The preference for texting over talking split the sample almost exactly in half. Of those reporting a preference, 370 stated that they preferred talking ("the Talkers") and 372 preferred texting ("the Texters"). However, Texters were significantly younger (mean $(M) = 22.7$ years, standard deviation $(SD) = 6.3$) than Talkers ($M = 24.3$, $SD = 7.7$, $t(695) = 3.2$, $p < .01$), and were more likely to be female (χ^2 (1, $n = 740) = 25.1$, $p < .01$).

6.3.2 How Texters and Talkers Use Their Mobiles

Text messaging emerged as a qualitatively different pattern of mobile phone usage in the present sample. The results presented in Table 6.1 indicate where the main differences in usage lie. Although Texters ran up significantly lower monthly bills than Talkers, they spent more of it on texting, sending nearly twice as many texts per month as Talkers, and under half the number of voice calls. Texters were also more likely to have a special tariff for texting and were usually on pay-as-you-go as opposed to contract agreements. Clearly, their choice of pricing plan was adjusted to suit their texting needs and preferences. Texters also committed more time and effort to the process of texting, they were more likely than Talkers to edit and rewrite their text messages, and their messages were more likely to make use of the full character limit available on their mobile phones. Nearly half of the Texters believed they texted "too much"; conversely, Talkers were more likely than Texters to receive more texts than they sent.

Table 6.1 *t*-test and χ^2-test for how Texters and Talkers use their mobiles

Questionnaire item	M	t
Monthly bill (£) (*df* = 617)		
Talkers	32.96	3.2***
Texters	26.43	
Amount of bill that is text (£) (*df* = 550)		
Talkers	8.58	−5.2***
Texters	14.06	
Number of texts sent per month (*df* = 609)		
Talkers	120.3	−6.4***
Texters	231.6	
Number of voice calls made per month (*df* = 428)		
Talkers	92.9	6.2***
Texters	41.1	
	Percentage	χ^2
Special tariff for texting (*n* = 742)		
Talkers	43.2	7.4**
Texters	53.2	
On Contract (*n* = 742)		
Talkers	66.2	35.9***
Texters	44.4	
Receive more texts than send (*n* = 742)		
Talkers	57.8	5.9*
Texters	48.9	
Text too much (*n* = 742)		
Talkers	22.4	9.96**
Texters	48.8	
How often write/rewrite tests (*n* = 336)		
Talkers (*often/sometimes*)	39.2	26.3***
Texters (*often/sometimes*)	48.0	
Phone use changed over time (*n* = 742)		
Talkers	56.5	0.05
Texters	55.6	
Length of texts (*n* = 655)		
Talkers		
All limit	39.4	
Most	32.9	
Less than half	27.7	21.8***
Texters		
All limit	47.9	
Most	37.3	
Less than half	14.8	

*$p < .05$; **$p < .01$; ***$p < .001$.

About half of the sample reported changing the way they used their mobiles over time. However, Texters did not differ from Talkers in this, suggesting that the preference for texting or talking is not dependent on changes of use occurring over the length of ownership. It also suggests that the preference for texting does not progress through stages, but arises as a preference for this distinctive medium of communication in its own right.

6.3.3 What Texters Get Out of Texting

This preliminary analysis suggests some clear distinctions, both quantitative and qualitative, between the way Talkers and Texters use their mobile phones. In this section, the focus is on what it is about texting and talking that motivates mobile phone owners to use their mobiles in these distinctive ways. In particular, how mobile use has changed the quality of respondents' personal relationships, and the extent to which they feel able to express themselves through texting and voice calls, are explored.

Table 6.2 reveals that Texters had deeper relationships with the person they texted most than did Talkers, for example, by responding positively to items, such as "I feel quite close to this person", and negatively to items such as "I would never tell this person anything intimate about myself". However, Texters and Talkers did not differ in the reported breadth of their relationships, as probed through questions such as "We contact each other in a variety of ways beside text" and "Our text messages are limited to just a few specific topics". Possibly as a result of this greater depth of personal contact, Texters were significantly more likely to report that texting had affected their relationships with friends and family. They also reported that texting helped them develop new relationships, as well as add something new to their existing relationships.

Texting also appeared to alter the way that Texters expressed themselves. Whilst most Texters and Talkers felt that face-to-face contact was the best medium for expressing themselves, more than twice as many Texters preferred texting for this purpose than Talkers preferred voice calls. Texters seemed more able to express their "real selves" through text messages, not only were they more comfortable saying things in texts than face-to-face, but they also felt their family would be surprised if they were to read the content of their messages. Text messaging seems to afford Texters an opportunity for more intimate social contact than it does Talkers.

6.3.4 The Social Ecology of Texting: Textmates and Text Circles

The distinctive affordances of texting that attract Texters appear to create a special kind of "text world" with its own social ecology and structure. This section reports answers to a group of questions that reveal something about

Table 6.2 *t*-test and χ^2-test on what Texters get out of texting

Questionnaire item	*M*	*t*
Depth of relationship with closest textmate (*df* = 563)		
Talkers	28.8	−3.5**
Texters	30.7	
Breadth of relationship with closest textmate (*df* = 562)		
Talkers	14.37	−1.6
Texters	14.93	
	Percentage	χ^2
Extent to which family surprised if read texts (*n* = 742)		
Talkers		
Not at all	72.7	
Slightly/very	27.3	23.6***
Texters		
Not at all	56.2	
Slightly/very	43.8	
Medium feel better expressing self (*n* = 708)		
Talkers		
Text	7.6	
Voice	11.5	
FTF	80.8	52.5***
Texters		
Text	28.2	
Voice	6.7	
FTF	64.9	
More comfortable saying things in text than FTF (*n* = 742)		
Talkers	23.2	24.2***
Texters	40.1	
Texting affected relations with friends (*n* = 598)		
Talkers	26.0	25.1***
Texters	45.6	
Texting affected relations with family (*n* = 595)		
Talkers	9.4	14.1***
Texters	20.5	
Texting affected social life (*n* = 742)		
Talkers	38.3	22.2***
Texters	57.7	
Texting helped develop new relationships (*n* = 595)		
Talkers	29.3	15.6***
Texters	50.0	
Texting added new things to relationships (*n* = 592)		
Talkers	27.6	18.1***
Texters	45.0	
Text changed the way they express self (*n* = 589)		
Talkers	9.8	8.8**
Texters	18.4	

FTF: face-to-face; *p < .05; **p < .01; ***p < .001.

the extent and structure of the network of contacts maintained by Texters. The results for these questions are summarised in Table 6.3.

What emerges from this table is an indication of the social ecology of texting: Texters seem to establish and maintain social contacts within one or a few fairly well-defined and close knit groups of textmates, forming "text circles", within which they regularly, perhaps even continuously, exchange messages. Text circles may not be particularly extensive. Table 6.3 shows that Texters regularly text about the same number of people as do Talkers. However, they do have fewer contacts in their phonebooks than Talkers and these consist of more mobile numbers (essential to texting) than landline numbers. They also engage more frequently in extended "text conversations", sending nearly twice as many messages in these conversations as Talkers. Texters were more likely to text a particular group as opposed to many groups, and more frequently participated in several simultaneous text conversations, findings which taken together reinforce the idea that Texters share interconnections within a close group of friends in perpetual text SMS contact with one another.

The questionnaire also probed for information about the phonebook contact that respondents texted most frequently, their closest textmate. Although Texters and Talkers did not differ in the length of time they had known or texted their closest textmate, or differ in the how often they met them face-to-face, Texters not only texted this person more frequently, they also phoned them less frequently. Furthermore, as Table 6.2 shows, Texters have significantly deeper relationships with their closest textmates than do Talkers; clearly these are special kinds of communicative relationship for which calls are no substitute.

6.3.5   Finally, Some Evidence for Social Compensation?

One of the key questions hoped to be explored with this survey concerned the underlying appeal of SMS text messaging for those who have established these close knit text circles.

Whilst we are still investigating this by analysing the reasons Texters and Talkers have provided for their preferences, it is possible to report some preliminary results that throw light on this. Specifically, it was found that Texters scored significantly higher on the Interaction Anxiousness Scale ($M = 41.3$, $SD = 9.9$) than Talkers ($M = 36.9$, $SD = 8.7$, t (728) $= -6.4$, $p < .001$), for example, rating items like "I wish I had more confidence in social situations" and "I often feel nervous in casual get-togethers" more strongly. Texters also scored significantly higher on the UCLA Loneliness Scale ($M = 21.3$, $SD = 5.4$) than Talkers ($M = 20.1$, $SD = 4.8$, t (730) $= -3.1$, $p < .01$). That is, they judged items like "How often do you feel you lack companionship?" and "How often do you feel isolated from others?" to be more characteristic of them. These scores seem at first sight to echo McKenna's observations concerning the appeal of the Internet to the lonely

Table 6.3 *t*-test and χ^2-test on textmates and text circles

Questionnaire item	*M*	*t*
No. names in phonebook (*df* = 577)		
Talkers	84.3	2.1*
Texters	74.6	
No. people text regularly (*df* = 635)		
Talkers	12.8	−.31
Texters	13.2	
Average no.texts sent in text Conversation (*df* = 375)		
Talkers	4.2	−3.6**
Texters	7.4	
Questions relating to respondents closest textmate		
How long known this person (years) (*df* = 522)		
Talkers	4.4	0.45
Texters	4.2	
How long been texting this person (years) (*df* = 539)		
Talkers	1.9	1.66
Texters	1.7	
How often text person text most per month (*df* = 495)		
Talkers	60.1	−4.2***
Texters	94.6	
How often phone person text most per month (*df* = 544)		
Talkers	36.8	2.1*
Texters	29.4	
How often see this person FTF per month (*df* = 537)		
Talkers	16.2	−.29
Texters	16.6	

	Percentage	χ^2
Majority of phonebook are mobile numbers (*n* = 727)		
Talkers	83.5	15.3***
Texters	93.5	
Text particular group (*n* = 332)		
Talkers	67.1	4.2*
Texters	74.5	
How often participate in text conversations (*n* = 727)		
Talkers		
Never/rarely	48.6	
Sometimes/very often	51.4	66.1***
Texters		
Never/rarely	24.4	
Sometimes/very often	75.6	
Participate in multiple text conversations (*n* = 640)		
Talkers		
Never	49.5	
Rarely	31.2	
Sometimes/very often	19.2	
Texters		
Never	30.0	30.2***
Rarely	35.3	
Sometimes/very often	34.7	

FTF: face-to-face; *p < .05; **p < .01; ***p < .001.

and socially anxious: that computer-mediated contact provides a safe form of social engagement that compensates (possibly even substitutes) for direct face-to-face contact, particularly in the early stages in the formation of new relationships. However, whilst these results suggest that the socially anxious and lonely might get something out of texting that they cannot find in voice calls, they do not consistently point to this conclusion. Why this may be the case is explained in the following section.

6.4 Discussion and Conclusions

The results showed that there was a nearly equal split within the sample between those who preferred using their mobiles for talking and those who preferred texting. This distinction was not just a simple matter of preference: Texters were more likely to be younger and female, and differed from Talkers in the way they used their mobiles, their underlying motivations and in key aspects of their personality.

The finding that Texters tended to be younger and female seems to parallel the findings of the Pew Internet and American Life Project (2001). This was a three-year study that examined the role of the Internet in Americans' everyday lives. Gender and age differences were found in the respondents' use of the Internet to develop and maintain relationships, whereby young females were more likely to use and benefit from electronic communication in terms of relationships.

The fact that some people prefer texting to talking suggests that they get something out of texting that they cannot get from talking. Indeed, the results showed that texting affords a distinctive medium for personal contact. Not only did Texters report that the medium added something extra to their existing relationships with friends and family, but it also took them beyond this, helping them develop new relationships. As a result, they committed more time and effort to the process of message composition, writing longer messages and editing them more carefully, expressing things in their messages that they would not have felt comfortable saying face-to-face. Texters also developed deeper relationships with the person that they texted the most compared to Talkers, despite there being no difference in the amount of time they had both known the person, or had been texting the person. Texters did text this person significantly more than Talkers did however, and it is possible that this greater frequency of contact facilitated a deeper, more intimate relationship. However, it is not yet known whether Talkers show the mirror image effect with their voice calls, establishing deeper relationships with those they most frequently talk with, nor can there be confidence in the causal direction of this relationship until the questionnaire data are more fully analysed.

Further analysis of the data (Reid and Reid, under review) has shown that the preference for texting emerges most strongly among mobile owners who discover the affordances of texting for perpetual contact and especially for expressive communication. Therefore, it is probable that the preference for texting is a consequence of the relational benefits that accrue from texting, rather than a prior cause of them. Apart from its novelty or recreational value, it is also difficult to see why any preference for texting should arise in the absence of these benefits.

Perhaps the strongest theme to emerge from the data was the notion of "text circles". We believe that many Texters establish small, tightly-knit networks of textmates, with whom they exchange messages more or less continuously, engaging in extended text conversations consisting of multiple partners and multiple turns, even preferring this kind of contact over voice calls with members of their text circles. These findings neatly dovetail with Thurlow's (2003) observation that the majority of text messages fulfil phatic and social-relational functions, intended to share feelings, perform "friendship work" or establish a mood of sociability rather than to necessarily communicate information or to make practical arrangements. Texting then creates "a steady flow of banter (…) used (…) to maintain an atmosphere of intimacy and perpetual social contact. In this sense, text messaging is small talk par excellence – none of which is to say that it is either peripheral or unimportant" (Thurlow, 2003: 12). We are now following this up by exploring whether the messages exchanged by Talkers are primarily of the informational-practical kind.

These results at first sight lean towards the conclusion that texting has the greatest social impact on those mobile owners who are socially anxious and lonely, as might be predicted from McKenna et al.'s (2002) research. McKenna et al. reported that people high in social anxiety and loneliness felt better able to express their "real self" online as opposed to offline, eventually resulting in the formation of close "real-world" relationships that endure over time. The present survey shows a similar tendency for Texters to report being more comfortable expressing their real feelings in text messages than in voice calls or face-to-face contacts. Many reported their family would be surprised to read their text messages, suggesting that texting allowed Texters to present a self-image that differs from the one familiar to family members and others who know them well. Given the choice, the majority of both Talkers and Texters preferred to communicate face-to-face. However, nearly a third of our Texters preferred texting to face-to-face communication, compared to less than a tenth of the Talkers. So these results do suggest that texting does help to foster the formation of personal relationships that may not have otherwise have the chance to develop face-to-face.

A number of findings, however, do not fit neatly with this conclusion. Firstly, in the data there was no evidence for our hunch that textmate relationships were a staging post in the formation of deep and lasting friendships. Texters continued to text their closest textmates, even when they met them face-to-face on a regular basis. Furthermore, whilst texting may provide the psychological distance necessary to manage the self-image presented to new textmates, and the time and effort Texters devoted to message content and composition does suggest that these impressions are often carefully crafted, texting as a contact medium did not appear to slip into second place when the opportunity to make a voice call or meet face-to-face existed. In fact, Texters preferred to message their closest textmates even when they could call them. Furthermore, Texters appear to concentrate their messaging on a small number of relationships within well-defined text circles, whether or not these comprise new relationships. Texting therefore has the character of a distinctive and alternative mode of social contact which continues to appeal to Texters even when textmates are already close friends that they meet with on a regular face-to-face basis.

These are preliminary conclusions at this stage in the survey. Analysis of the reasons Texters and Talkers offer for preferring messaging in various relational contexts is planned, and will hopefully throw further light on the distinctive qualities of SMS text messaging as a social and expressive medium. In the meantime we believe our findings underscore yet again how readily users appropriate and co-opt new communication technologies (Carroll *et al.*, 2001) to meet their own imperatives; for example, in the present case the imperative to establish and maintain perpetual social contact.

References

AOL Mobile (2002) Text-messaging. Available from: http://www.mobile.aol.co.uk/redesign/gallery/flirt/

Bamrud J (2002) SMS: the human factor. Available from: http://www.thefeature.com/article.jsp?pageid=15434

Birnie SA, Horvath P (2002) Psychological predictors of Internet social communication. *Journal of Computer-Mediated Communication*, **7**. Available from: http://www.ascusc.org/jcmc/

Carroll J, Howard S, Vetere F, Peck J, Murphy J (2001) Identity, power and fragmentation in cyberspace: technology appropriation by young people. *Proceedings of the 12th Australasian Conference on Informational Systems*. Coffs Harbour, NSW, Australia, 5–7 December.

Chenault BG (1998) Developing personal and emotional relationships via computer-mediated communication. *CMC Magazine*. Available from: http://www.december.com/cmc/mag/

Danet B (1995) Playful expressivity and artfulness in computer-mediated communication. *Journal of Computer-Mediated Communication*, **1**. Available from: http://www.ascusc.org/jcmc/

Haig M (2002) Mobile marketing: the message revolution. Kogan Page Limited, London.

Kasesniemi EL, Rautiainen P (2002) Mobile culture of children and teenagers in Finland. In: Katz JE, Aakhus M (eds). *Perpetual Contact: Mobile Communication, Private Talk and Public Performance*. CUP Cambridge, pp. 170–192.

Katz JE, Aakhus M. (2002) Introduction: framing the issues. In: Katz JE, Aakhus M (eds). *Perpetual Contact: Mobile Communication, Private Talk and Public Performance*. CUP Cambridge, pp. 1–15.

Leary MR (1983) Social anxiousness: the construct and its measurement. *Journal of Personality Assessment*, **47**, pp. 66–75.

Ling R, Yttri B (2002) Hyper-co-ordination via mobile phones in Norway. In: Katz JE, Aakhus M (eds). *Perpetual Contact: Mobile Communication, Private Talk and Public Performance*. CUP Cambridge, pp. 139–167.

McKenna KYA, Bargh JA (1998) Coming out in the age of the Internet: identity "demarginalisation" through virtual group participation. *Journal of Personality and Social Psychology*, **75**, pp. 681–694.

McKenna KYA, Bargh JA (2000) Plan 9 from cyberspace: the implications of the Internet for personality and social psychology. *Journal of Personal and Social Psychology Review*, **4**, pp. 57–75.

McKenna KYA, Green AS, Gleason MEJ (2002) Relationship formation on the Internet: What's the big attraction?' *Journal of Social Issues*, **58**, pp. 9–31.

MDA (2004) 111 Million new year greetings sent. MDA. Available from: www.mda-mobiledata.org

Nokia (2001) 'Taking the show on the road – market study shows that 3G mobile users want to be entertained on the move. Nokia Press release. Available from: http://press.nokia.com/PR/2001

Nokia (2002) http://www.nokia.com

O'Brien K (1993) Improving survey questionnaires through focus groups. In: Morgan DL (ed.). *Successful Focus Groups*. Sage, London; pp.105–108.

Parks MR, Floyd K (1996) Making friends in cyberspace. *Journal of Personality Assessment*, **46**, pp. 80–97.

Pew Internet and American Life Project (2001) Teenage life online: the rise of the instant-message generation and the Internets impact on friendships and family relationships. Available from http://www.pewinternet.org/

Puro J (2002) Finland, a mobile culture. In: Katz JE, Aakhus M (eds). *Perpetual Contact: Mobile Communication, Private Talk and Public Performance*. CUP Cambridge, pp. 18–29.

Reid FJM, Reid DJ (under review) The hyperpersonal affordances of SMS: an Internet study of mobile phone text messaging. *British Journal of Social Psychology*.

Russell DW (1996) UCLA Loneliness Scale (version 3): reliability, validity and factor structure. *Journal of Personality Assessment*, **66**, pp. 20–40.

Thurlow C (2002) Generation Txt? Exposing the sociolinguistics of young peoples text-messaging. *Discourse Analysis Online*. Available from: http://extra.shu.ac.uk/daol/

7

Appropriating Tools and Shaping Activities: The Use of PDAs in the Workplace

Jenny Waycott

7.1 Introduction

This chapter looks at how handheld computers, or Personal Digital Assistants (PDAs), have been used as general-purpose workplace tools in two organisations. The chapter will examine the related processes of appropriating new tools and shaping existing activities through the use of new technologies. It will draw on interviews from each case study to examine how sociocultural factors influenced the way the PDA was appropriated as a workplace tool, and to identify how the PDA changed the activities it was used to support. Understanding these dual processes is particularly relevant in today's information technology (IT) environment, in which personal and portable technologies, such as mobile phones and PDAs, are becoming more and more pervasive. Such technologies are truly personal, in the sense that they often remain in the hands of the owner and can be used in many different ways. They also appear to inspire in users a sense of emotional attachment. Thus, understanding how new users appropriate mobile technologies, and how, in turn, those technologies change the way people do things, is a timely concern.

Tool appropriation is defined here as the integration of a new technology into the user's activities. The term "appropriation" has been chosen to signify that tool use is an active process. The user is not a passive recipient of the new technology but instead chooses to use it in various unique ways, adapting the use of the technology to suit the user's purposes. Wertsch (1998) defined appropriation as the process of "taking something that belongs to others and making it one's own" (Wertsch, 1998: 53). Similarly, Jennie Carroll and colleagues described it as a transition from "technology-as-designed" into "technology-in-use" (Carroll *et al.*, 2002b). In other words, by appropriating a technology the user exerts ownership over it, adapting the new tool so that it can be successfully integrated into the user's social settings.

This chapter will draw on concepts from activity theory to examine tool appropriation and the shaping of activities through the use of new tools. Activity theory is a sociocultural framework emanating from Soviet psychology and Marxist philosophy (e.g. Leont'ev, 1978; Vygotsky, 1978), which has recently received much interest in the human–computer interaction (HCI) and computer-supported cooperative work (CSCW) research communities (e.g. Kaptelinin, 1996; Nardi, 1996; Turner and Turner, 2001; Halloran et al., 2002). The main advantage of activity theory to these research fields is that it offers a way of examining the use of computer tools in context, taking into account the goals of the user and the social settings in which the user's activities take place. Activity theory is not a theory as such in that it is not "a fixed body of accurately defined statements" (Kuutti, 1996: 25). Rather, it is a collection of broadly defined concepts that are open to interpretation. This chapter will focus on the following central notions, which have been selected because they are particularly relevant to this research:

1. Activities are sociocultural systems.
2. Activities can be divided into hierarchical layers of actions and operations.
3. All activities are in a constant state of transformation, involving the introduction and resolution of contradictions in the activity system, and breakdowns in the actions and operations that make up the activity.

The following pages describe each of these concepts in further detail. Meanwhile, the next section will briefly introduce each of the two case studies upon which this discussion is based. This will be followed by an examination of how sociocultural factors, as identified by an activity theory analysis, appeared to influence the way the PDAs were appropriated in each case study setting. Finally, the chapter will turn to an examination of how the PDA appeared to change the workplace activities it was used to support.

7.2 The Studies

The case studies described in this paper were carried out as part of a wider evaluation of PDA use in different learning and workplace settings (see Waycott, 2002; Waycott and Kukulska-Hulme, 2003). PDAs are palmtop computers, traditionally used as business tools to support time and information management (Geisler, 2003). For example, they can be used as electronic diaries, address books and notepads. In both studies described here, these were the functions that proved to be most useful to participants.

7.2.1 Institute of Educational Technology Case Study

In the first study, which took place over two years, 11 academic staff members in a university department were given a selection of PDAs to use as general workplace tools. These included Palm, Handspring and Hewlett Packard Jornada PDAs. However, the purpose of this study was not to compare these devices, nor to evaluate the PDAs against a set of usability criteria. Rather, the aim was to explore how well the PDA supported different work activities and how, over time, the new tool came to be integrated into users' activity systems. This study took place in an academic department, the Open University's Institute of Educational Technology (IET), which conducts research into the use of technology in education. Participants in this study were therefore likely to keep abreast of new technologies. All participants were interested in exploring the ways in which PDAs could support their work and learning activities. They expressed enthusiasm about testing out different features and functions of the device and were keen to explore the different activities for which the PDAs could be used. However, as described below, this did not mean that all participants used the PDA extensively; instead, participants varied greatly in the way they responded to and used the new device.

7.2.2 NatGasCorp Case Study

The second study examined PDA use in a large international organisation, known here as NatGasCorp (NGC). NGC explores and markets natural gas resources around the world. The organisation was originally part of a parent company; following a demerger in October 2000 it became a distinct company which deals with the development of international business assets. It operates in various countries, including Brazil, Argentina, the Philippines, Trinidad and Tobago, India, Egypt, Tunisia and Kazakhstan. The corporation's headquarters are based in the UK and many of the staff who work there are required to travel frequently to visit the assets and other overseas businesses. For this case study, 16 staff members were interviewed, most of whom travel frequently and extensively as part of their job. Therefore, this organisation provided a rich context in which to examine mobile working practices and the use of mobile technologies.

PDAs were introduced to this organisation three years prior to the study. It was initially hoped that the PDAs would be valuable tools for mobile workers to maintain access to the company Intranet. Palm Vx PDAs were supplied to managers in 1999 and the company's Intranet site was made available on the Palms through AvantGo software. The purpose of the Intranet was to provide a facility for storing company information, such as press releases, company policies and staff details. It was hoped that having this information available on PDAs would make it more accessible for staff who frequently worked away from the head office. However, at the time that the research was conducted, the Palm PDAs were no longer being used for this purpose. Although some of the staff still used PDAs, very few people used them to download the company Intranet, and most used them simply as electronic diaries and address books. The company's IT team was concerned about whether this was an efficient use of the PDAs, and so they were in the process of replacing the Palms with other "better value" devices.

7.2.3 Overview of Findings

Both studies primarily involved interviews with participants, which were supplemented by occasional observations of work practices. One of the salient findings was that participants differed greatly in the way they responded to and used the PDA. In NGC, some participants felt that they were making "inefficient" use of the device (e.g. "I do not use it efficiently at all. It is just used for meetings and addresses"), while others disagreed with the notion that they were not getting value out of the PDAs. For these people, using PDAs as diaries and address books, or simply as a way of synchronising information between work and home computers, did provide support for their working practices. Meanwhile, there were some participants who used the PDA more extensively (e.g. downloading e-mails and documents onto the PDA, note-taking, managing expenses) and some staff who no longer used the PDA at all.

Similarly, participants in IET varied in the extent to which they used the PDA. Some found the PDA to be a very useful workplace tool and used it extensively throughout the study. One participant did not use the PDA at all, finding that it did not offer sufficient benefits over other tools. Meanwhile, some participants were initially enthusiastic about the device but found that over time, its limitations outweighed its benefits and it no longer served their needs as well as it had originally done.

In order to understand this outcome, the data collected in each study were examined to answer the following question: "Why did participants vary so greatly in the way they appropriated the PDA?" Activity theory concepts were useful in making sense of the data and addressing this question. In particular, Engestrom's (1987) activity system framework was used to describe tool appropriation as an activity in itself, providing a way of identifying the socio-cultural factors that appeared to exert an influence over the process of tool appropriation. This adaptation of Engestrom's model has been labelled the "Activity System Tool Appropriation Model" (ASTAM), and is introduced below, following an overview of Engestrom's depiction of activity systems.

7.3 Activity Systems

In activity theory terms, an activity is defined as a social system. In this sense, activities involve communities of people and are embedded within particular social settings. For instance, "running" becomes not just running, but "running

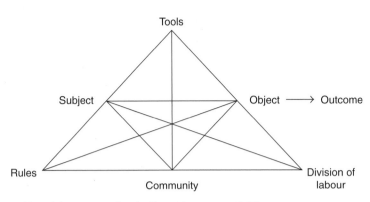

Figure 7.1 The activity system triangle (from Engestrom, 1987).

in a race", incorporating all of the rules, tools and established practices that this implies. This idea is typically illustrated by Engestrom's (1987) activity system triangle (Figure 7.1). This framework incorporates the subject, or person acting on the world, the object towards which the activity is directed (representing the activity's motive) and the tools, through which the activity is mediated. Engestrom added to this basic triangle three further components: the community, rules/regulations and division of labour, which each have a mediating role in the execution and development of the activity. Members of the community are involved in activities that are related, but not identical, to the central activity system. The rules or regulations in an activity system can consist of both formal laws and procedures as well as more informal and implicit ways of doing things. Rules share many similarities with the concept of tools and it may sometimes be a matter of interpretation as to whether a particular artefact is classified as a "tool" or a "rule". The division of labour "refers to both the horizontal division of tasks between the members of the community and to the vertical division of power and status" (Engestrom, 1993: 67). It is through the horizontal division of tasks that actions are divided among members of the community, which together help the subject to reach the objective of the activity.

7.4 The Activity System Tool Appropriation Model (ASTAM)

Activity theory is clearly a relevant framework for research investigating the impact of new technologies on social practices. One of the central tenets of activity theory is the notion that all activities are mediated by the use of tools, both conceptual tools (such as language) and physical tools (such as technological artefacts) (Leont'ev, 1978; Vygotsky, 1978; Nardi, 1996). The emphasis on tools as mediators of activity focuses attention on the activity itself rather than simply the interaction between the human and the computer. In other words, the human is seen to be doing something other

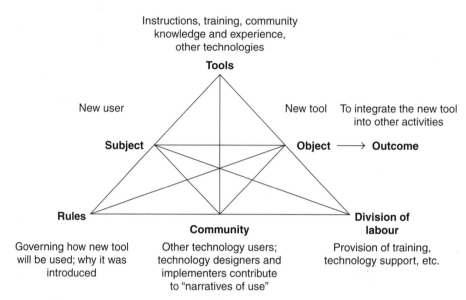

Instructions, training, community
knowledge and experience,
other technologies

Tools

New user   New tool   To integrate the new tool
into other activities

Subject   **Object** ⟶ **Outcome**

Rules   **Division of**
Community   **labour**

Governing how new tool   Other technology users;   Provision of training,
will be used; why it was   technology designers and   technology support, etc.
introduced   implementers contribute
to "narratives of use"

Figure 7.2   The ASTAM.

than using the computer: the computer is the tool through which the user achieves objectives. However, as Bodker (1996) acknowledged, artefacts not only mediate our interaction with the world; they can also be the objects towards which our activities are directed. Drawing on this notion, this chapter uses the activity system framework to analyse the process of tool appropriation as an activity itself, where the object of the activity is the tool being appropriated; in this case, the PDA. The mediating artefacts in this activity include both the tools used to support the appropriation activity (for instance, instruction manuals specifying how to use the new tool) and the existing artefacts with which the new tool must become integrated. The community includes other people who are also involved, to some extent, in the activity; for instance, colleagues, friends and technology providers. The rules and division of labour might include regulations governing the use of the new tool (for instance, how it is expected to be used) and a separation of responsibilities for installing software and overcoming technical difficulties. This model is illustrated in Figure 7.2. In the following discussion each of the components in the ASTAM framework will be examined in turn, focusing on how these factors appeared to influence the way the PDA was appropriated in each of the two case study settings described above.

7.4.1 The Subject

In both studies participants brought to the tool appropriation activity their own unique combination of past experience with other technologies, personal inclination towards using new technologies, and personal preferences for different work and study practices. These each appeared to have an effect on how the PDA was used. This was particularly apparent with respect to how people adapted to using the new limited text input methods on the PDA, especially the handwriting recognition systems, such as Graffiti on the Palms (for more information about Graffiti, see Sears and Arora, 2002). A common basis of comparison was touch-typing on a computer keyboard. Those who were touch-typists compared handwriting on the PDA unfavourably with typing on a full-sized keyboard. Meanwhile, those who did not like typing generally responded more positively to the handwriting recognition systems on the PDA.

Participants also differed in their general working practices. For example, one participant in IET said that she had always been a very organised person and this had an impact on the way she used the PDA, particularly with respect to using its diary function. Previously, when using a paper diary, she had been "very, very particular" about it and made sure it adequately supported her work activities. She has adopted a similar approach to using the PDA, keeping the diary up to date and ensuring it is frequently synchronised. Another interviewee described how she had previously used a diary as both a time and information management system, keeping printouts of appointment schedules, agendas and e-mail messages in her paper Filofax folder. The PDA fitted neatly into this preferred way of working. Instead of keeping printouts, which were awkward and messy to maintain, the PDA enabled her to keep electronic copies of meeting notes and agendas, which could be attached to her appointment schedule. She did not have to modify her preferred way of working, but the PDA provided a more efficient means of achieving this.

One interviewee from NGC described the disparity between the need to match tools with users' personal preferences and the predilection within organisations to provide users with standardised technologies and software: "IT groups tend to want standardisation, same software, because it's much easier to service it, costs are low and they're in control. Users, on the other hand, want something they're comfortable with … And that's what IT should be working towards so that people can have the choice that makes sense for them and they provide the interconnectivity. Because the expensive part of it is in the interconnectivity, it's not in the individual devices … It's almost like a piece of stationery. And to have someone say you need to have stationery that has a blue binder with this much spacing on the lines, is dealing with concepts that are no longer really current."

This quote suggests that there is a general contradiction between the needs of the subject in the activity system (the technology user) and the concerns of other members of the community (the technology providers). This interviewee also likened the PDA to a piece of stationery, which is a useful analogy. It calls to mind the tool history of the PDA. As Geisler (2003) described, precursors of the PDA included paper diaries and information management systems, such as the Filofax and Rolodex. These tools have traditionally been viewed as stationery items in the paper-based office. Understanding this tool history provides insight into why the PDA is primarily used as a diary and address book, which may displace the notion that using the PDA for these functions is an inefficient use of the device.

7.4.2 Tools

A recurrent theme in the interviews was that the PDA was not a stand-alone tool. It was used alongside other technologies and as such the process of appropriating the PDA involved integrating the new device with the tools that were already used. In order for the PDA to be successfully adopted and used, it was important that it complemented, rather than conflicted with, existing tools. Participants in both studies used the PDA as an adjunct to their main computers. The PDA provided a way of accessing or creating information in digital format that previously could only be dealt with on a desktop or laptop computer. The new tool, then, made it possible to access this information in situations where a computer would not typically be available, thus extending the capabilities of existing tools.

However, in some cases the PDA appeared to be a redundant mobile technology, particularly for those who always had access to a laptop computer. This was especially the case for staff at NGC who considered themselves to be active mobile workers. The division of labour within NGC determined whether a staff member was an active or passive mobile worker. Those whose jobs involved, for example, collecting and analysing data, writing notes on their findings, producing reports and communicating the results of their work to colleagues, were active mobile workers. For these people, a laptop computer was an essential mobile working tool and was particularly important for those using sophisticated technical software. For example, a core group in the organisation was the petrophysics teams; petrophysicists were required to travel to gas and oil exploration sites to assess the findings at those sites. Their activities were supported by the use of specialist technical software, used to calculate data and produce large graphical charts of their analyses. When working remotely, petrophysicists used this software on a laptop computer. Similarly, employees in the Risk Analysis team travelled to overseas assets to carry out analyses of the business risks at those assets. The technical tools used to support this activity included word processing and presentation software for preparing written and verbal reports, the Lotus Notes system for sharing documentation, and e-mail for collaborating with colleagues. For that reason, all Risk Analysis officers used a laptop computer when working remotely.

In such cases, the laptop computer would always be available to the worker and some functions that could be performed on the PDA could be more easily and effectively carried out on the laptop. Therefore, there was a conflict between the two tools and there was little incentive for such workers to use the PDA. However, other staff members, whose jobs primarily involved overseeing the work of team members or attending meetings outside of the office, could feasibly replace a laptop with a PDA. For these people it was more important to have access to information, such as their calendar and contacts database, than to have the ability to produce reports using a computer with a full-size keyboard.

7.4.3 Community

In both studies, participants made use of knowledge elicited from the communities to which they belonged. Members of the workplace community, as well as the wider community of friends and family, were an important source of knowledge about how the new tools could be used and integrated into work activities. Community knowledge was also useful for providing trouble-shooting strategies when technical difficulties arose. Many participants from IET acknowledged the support of IT staff in the workplace who helped overcome technical difficulties and provided support for the process of setting up the device and ensuring it was successfully integrated with other workplace tools.

Furthermore, some participants spoke of how they had witnessed other members of the community using different PDAs. This served to highlight benefits and limitations of the devices colleagues were using, and also provided information about how PDAs could be further exploited to support workplace activities. Community knowledge also played a part in one IET interviewee's decision not to integrate the PDA into his workplace activities. He saw that other people found the data input methods irritating and this confirmed his belief that the tool would be difficult to use. In addition, he read articles in the press that also highlighted limitations of the device.

This is an example of what Churchill and Wakeford (2002) termed "narratives of use": general discourses relating to how technologies (in particular, mobile technologies) can and should be used. They argued that such narratives are gleaned from media representations, such as advertisements and newspaper reports, as well as from "personal experiences and the experiences of others one knows or can observe" (Churchill and Wakeford, 2002: 164).

In NGC there was a strong "discourse of mobility" regarding how the PDA should be used. Some participants felt that their use of the PDA was inefficient, reflecting the IT team's belief that employees were not getting best value out of the PDAs. However, an examination of interview comments revealed that PDAs had actually been beneficial to some participants, particularly as a tool for supporting time management. This suggests that the IT team's decision to replace the Palm PDAs with higher-specification devices is based on assumptions of how such devices should be used, rather than a full understanding of staff work activities and how the PDAs fit in to those activities.

7.4.4 Rules and Division of Labour

The community also contributed to the tool appropriation activity through the rules and division of labour that mediated the activity. For instance, in IET, departmental rules and informal regulations were called upon when participants encountered technical problems with the PDA. It was standard practice to ask for help from the department technical support team when such problems occurred. The task of resolving technical problems was then divided among members of the IT support team, the PDA user and other members of the community who were able to contribute appropriate expertise.

In some instances, conflicts arose between the PDA user and the IT support team; there was a contradiction between the PDA user's objectives and the IT team's objectives. The activities of appropriating the PDA and providing IT support were co-existing: they took place within the same social context and when device breakdowns occurred, the activity of PDA appropriation became dependent on the activity of IT support provision. A contradiction occurred because the IT support activity had its own set of rules (i.e., priorities for providing support to various projects and staff members), division of labour (the availability of support staff to help resolve problems) and objectives (to provide technical support across the entire department). These were different from the rules, division of labour and objectives of the PDA activity; that is, to integrate the PDA effectively into work practices. Consequently, there was a contradiction between the two activities, which had the effect of disrupting the tool appropriation activity, making it difficult at times for participants to overcome technical difficulties and continue using the PDA to the extent that they wished.

7.4.5 The Object

The tool itself, the object towards which the activity was directed, presented new possibilities and constraints. These had an effect on the way the PDA was appropriated. For instance, battery life limitations meant that some participants had to modify the way they used the PDA in order to avoid having to constantly replace or recharge batteries. Meanwhile, some participants adapted the way they used the PDA in order to overcome the text input limitations, using the PDA in conjunction with a foldout keyboard, or relying upon abbreviations and shortcuts when entering text using the handwriting recognition system.

In contrast, the possibilities of the PDA – primarily, its portability – gave participants the opportunity to enhance their work activities. The PDA provided a means of storing information in electronic format in a small, lightweight device that could fit in the palm of one's hand. Furthermore, the PDA could be easily synchronised with the desktop or laptop computer, giving portable access to data normally stored on the main computer: "I see the PDA as taking bits of my computer with me when I go".

One outcome of the successful appropriation of the new tool was an increased dependence on the PDA. A common view among those participants who had successfully integrated the PDA into their activities, was that the new tool was their "friend" and that they would feel lost without it: "It goes in my handbag with all those other crucially important things like my credit card, my reading glasses, my car keys and my PDA. Those are what I need to function".

This outcome may be particular to the PDA, which is highly portable and therefore affords emotional attachment by being constantly available. Its portability means that it can always be carried on the user's person, either in a shirt pocket or handbag. Thus, it is a truly personal device that commands emotional dependence. Whether this aspect of tool appropriation is common to all technologies, however, remains to be seen.

Furthermore, this emotional dependence was not so apparent in NGC. This could have been due to the timing of the study, which took place three years after PDAs had been introduced in NGC. That is, the enthusiasm for the PDA among NGC staff may have waned over the three years since the tools were introduced. This also appeared to occur over the two years of the IET study; that is, some participants became less enthusiastic about the device and stopped using it as extensively as they had initially done. This is in line with Jennie Carroll and colleagues' depiction of technology appropriation (Carroll *et al.*, 2002a, b). They argued that technology appropriation is not stable; once a technology has been appropriated, it is continually evaluated, reassessed and updated. Use of a particular technology, then, may decrease over time, as it becomes replaced by other technologies or if it no longer adequately serves the needs of its user. After all, users' needs also continually evolve.

This discussion has demonstrated that there are multiple interacting influences over how users appropriate new technologies. The ASTAM framework has been a useful way of identifying and framing these influences. However, this only tells us one side of the story. While various elements within the social setting in which a technology is used clearly help to shape how that tool is used, the tool itself can also be responsible for changing that social setting. New tools can change activities in many, sometimes surprising, ways (Carroll *et al.*, 1991). This aspect of PDA use in the two case studies is discussed further below. First, though, the activity theory concepts that help to explain how new tools change activities will be briefly introduced below.

7.5 Tools Shaping Activities

One of the central tenets of activity theory is the notion that all human activities are mediated by the use of tools, both conceptual tools (such as language) and physical tools (such as technical artefacts). The term mediation refers to how objectives are achieved through the use of tools. However, when used as mediating tools, technologies have also the power to change activities. Therefore, understanding how tools mediate activities is closely aligned with the more general concern of understanding how technologies shape society, that is, how technologies change the activities they are used to support.

The activity theory concepts of contradictions, actions and operations are particularly relevant to understanding how new tools change activities. According to Engestrom (1993), all activity systems are in a constant state of development and transformation. This is in part due to contradictions, which occur both within and between activities. It is through the introduction and resolution of contradictions that activity systems evolve. Contradictions "manifest themselves as problems, ruptures, breakdowns, clashes" (Kuutti, 1996: 34). There are always contradictions in an activity system and they are necessary, although disruptive, for the development of the activity.

Activities can also be modified by the actors (subjects) involved in the activity, who develop and introduce new mediating tools, create and change rules and regulations, redistribute the division of labour and so on. Changes may be apparent within the social context of the activity, such as the community, rules and division of labour, and they may also appear within the actions and operations that make up the activity. Actions and operations are not represented in Engestrom's extended triangle model, which is concerned with the social context of activity rather than the mechanisms by which the activity is actually carried out. Actions and operations can be conceived of as hierarchical layers, which provide the means by which the object of the activity is transformed into an outcome (Leont'ev, 1978). The operations are the routinised processes that are carried out to perform the action. The action is a conscious process with a specific goal that helps to meet the overall objective of the activity. An example is the activity of work-related time management. Actions that contribute to this activity might include scheduling a new meeting appointment with colleagues or carrying out a work plan for the year ahead. The operations through which these actions are realised would include discussing the meeting appointment with colleagues, writing the appointment in a diary, checking to see what events and deadlines were scheduled for the year and recording goals or deadlines on a calendar.

Like all elements of an activity system, actions and operations are not static but are under constant development. For instance, some actions may, over time, become routine processes that no longer require conscious awareness. Thus, they become operationalised; that is, transformed into an operation. In this case, the new operation would be used to execute other goal-directed actions. The often cited example that Leont'ev (1978) gave was that of learning to drive a car, in which the process of changing gears begins as a conscious goal-directed action but, with time and practice, becomes an operation.

Conversely, operations can become actions when the conditions of an activity change. For instance, changing gears might be raised to the level of conscious awareness again; that is, become an action, when a new car is used.

As the driver becomes familiar with the gear-box on the new car, the action of changing gears would again be operationalised.

Bodker (1991) gave the example of using a new word processor when writing a letter. In this instance, the operations that had been built up through the use of more familiar tools would no longer apply and the user would need to execute each process involved in writing the letter as a specific goal-directed action. This can be likened to any instance when a new computer program is used to support a work activity. Initially the "repertoire of operations" at the user's disposal cannot be applied to the new tool (Bodker, 1991: 27). However, as the new tool becomes familiar, users may develop a new set of operations that relate specifically to the use of that tool. As more operations relating to that tool develop, use of the tool becomes easier and less prone to breakdowns.

7.6 How the PDA Shaped Workplace Activities

The following discussion draws primarily on data from the first case study which examined the introduction and use of PDAs in IET over a two-year period. Inevitably, the PDA changed the activities it was used to support, and the activity theory concepts described above are useful for explaining these changes. For instance, the PDA both introduced and resolved contradictions in the users' activity systems, leading to the development of those activities. Furthermore, the PDA had a disruptive effect on the actions and operations that made up the users' activities, typically making existing operations obsolete and requiring new tool-oriented actions. That is, users had to focus on the PDA itself in order to do tasks such as write meeting appointments in the diary or take notes on the memo pad.

In IET, the PDA was used most extensively to support the general workplace activities of time and information management. These activities involved tasks, such as keeping a diary, setting up meetings, doing yearly work plans, recording meeting notes, filing those notes or incorporating them into a document, reading and writing e-mails, keeping track of relevant information resources (such as newspaper articles and web sites) and recording and storing reminders to oneself (such as to-do lists). In IET, the PDA was most commonly used to support diary management, reading e-mails and taking notes; these are each discussed in turn below.

7.6.1 Diary Management

Prior to receiving the PDA, some participants already used an electronic diary, the Outlook calendar, on their desktop or laptop computers. Using the Outlook calendar meant they could share their schedule information with colleagues; for example, they could make the diary accessible to secretaries or to other team members. Furthermore, the diary could be used in conjunction with the e-mail system so available appointment times could be e-mailed directly to colleagues when meetings were being arranged and the appointments would be automatically recorded in the calendar. However, a major limitation of this system was its lack of portability: the electronic diary could only be accessed while the user was in the office or logged on to the network from home. For this reason, some participants had previously resisted using the Outlook calendar, preferring instead to use a paper diary. For these participants, the PDA provided the impetus to change to an electronic diary.

The decision to use the Outlook calendar had a great impact on the activity of time management. This could be seen primarily in the effect the new tool had on the community and division of labour within the activity. The time management activity took place within a community involving colleagues with whom participants made appointments and secretaries who often had responsibility for managing the academics' diaries. Keeping a diary in electronic format changed the division of labour within this community. For example, one participant found that by using an electronic diary, which colleagues could view on their desktop computers, he was able to reduce his involvement in the process of arranging appointments: "I seem to be spending more and more of my time making appointments with people and if I can just say go away and you can sort it out and tell me when you're free to fit in with me, I don't have to reply to emails saying can you give me your availability for December, January and February. [...] [The PDA] enabled me to store my diary electronically and that's slowly reducing my workload because I don't have to tell people when I'm available, I can say look at my electronic diary. I have had some meetings made electronically and that's been useful."

For those who already used the Outlook calendar, the PDA helped overcome a contradiction within the activity system caused by the tool's lack of portability. Previously, the Outlook calendar could only be accessed on the desktop or laptop computer, therefore those who relied on this system rarely had access to their schedule information when attending meetings in other offices. Some participants described strategies they would use to overcome this limitation. For example, they might keep a separate paper diary, print out pages from the Outlook calendar, or simply rely on their memory of the information contained within the electronic diary and make tentative appointments based on that memory.

There was a contradiction, therefore, within the tools that mediated this activity. The electronic diary on the desktop computer was beneficial for some purposes, but it could not be used outside the office. For some people, this meant using two different diaries: paper and the computer diary; with all the problems that this entailed, such as keeping the information in the two systems up to date and having to record new meeting appointments twice. There was always a danger, therefore, of double-booking.

Although the PDA was clearly useful as a time management tool and helped to resolve this contradiction, it also introduced some limitations due to its small screen size and awkward data input methods. The amount of information that could be displayed on the PDA diary was, of course, quite limited compared to how much could be viewed on a desktop computer or a paper diary. This meant that participants were unable to undertake actions that they were able to do easily on a paper diary, such as year planning. Viewing the diary on the PDA was quite different to looking at a paper diary and this required some adjustment: "I feel that I don't visualise it in the same way as I used to and that may be to do with the way the diary is laid out on the page."

Furthermore, awkward data input methods meant that, for some participants, previously operationalised processes, such as entering an appointment in the diary, became actions that required individual attention. Each action was more time-consuming and required greater effort than it would have done had it been operationalised. Therefore, some participants found it difficult to keep up with other people when entering appointments in meetings, and they had to devise strategies to overcome these difficulties. For example, they would use abbreviated words and ignore punctuation, or rely on other tools, such as pen and paper, to note down meeting times, which would later be entered into the electronic diary on the desktop computer. Thus, using the PDA diary did not entirely eradicate the need to carry paper or to enter appointments twice, at least for those who chose to appropriate the PDA in this way.

7.6.2 E-mail

Some participants in both studies found the PDA to be quite useful for downloading e-mails from the desktop computer, which they could then read while out of the office. They were not able to download e-mails "live" while on the move, as the PDAs did not have the capability of wireless connectivity. However, they were able to download the current contents of the e-mail in-box when synchronising the PDA with the desktop computer. This meant that e-mail use was no longer confined to an office environment. The PDA could be used to read and delete e-mails when the user had some spare time; for example, while on a train, while waiting for a doctor's appointment, or during a meeting or conference when the user was not involved in the discussion. This helped to resolve a contradiction within the activity system between the rules governing e-mail communication and the tools used for accessing e-mails, primarily the desktop or laptop computer.

That is, in IET there appeared to be a general unwritten rule that much communication and information sharing among colleagues would be conducted via e-mail. This was particularly the case for those who were undertaking collaborative work with colleagues not based at the university, or with colleagues who were absent due to study leave or away at a conference. Therefore, participants in this study managed quite large e-mail in-boxes, and they tended to check their e-mails on a daily basis when possible. Many of the e-mails they received, however, were not specifically relevant to them; these included housekeeping e-mails that were sent around the department or university, e-mails that derived from discussion lists and unsolicited e-mails from unknown sources.

Included among these irrelevant or less urgent e-mails, were some that were directed to the specific person, sometimes requiring urgent action or attention. For example, there might be an e-mail telling the recipient of changes to a meeting scheduled for the following day, or an e-mail from a colleague about a collaborative paper that needed to be completed in time to meet an imminent deadline. Therefore, the process of reading e-mails involved a combination of checking, filing or deleting non-urgent and irrelevant e-mails, while seeking out messages that required urgent attention or follow-up action.

The workplace desktop computer was the main tool used to access e-mail, although some participants also used home computers and laptops, and in some instances it was possible to check e-mails in specially equipped computer rooms when attending conferences. In each of these cases, however, the user was confined to a particular setting while undertaking the sometimes arduous task of managing the e-mail in-box. The PDA helped to overcome this contradiction: "I was able to scan through my e-mails at the end of the day … Either at home or on the train to a meeting the next day. So while out of the office, I would find if there was anything urgent then I would deal with it as early as I could."

However, one limitation of the PDA as a tool for reading e-mails was that it did not retain the file structure of the in-box on users' desktop computers, making it impossible to transfer e-mails to personal folders. Some participants managed their e-mail in-boxes, which were subject to a strict size limitation, by filing e-mails into different folders that had been created for this purpose. In other words, e-mails would be removed from the in-box and categorised according to user-defined categories. The PDA, however, only downloaded the main in-box, which meant that e-mails that had already been filed were not available to read on the PDA. Furthermore, it meant that the process of filing e-mails was an action that could only be undertaken using the desktop or laptop computer. For this reason, some participants chose not to use the PDA to read and manage e-mails.

Furthermore, space-saving mechanisms on the PDA meant that long e-mail messages were truncated and attachments could not be opened. This was particularly a problem for those participants whose jobs involved reading numerous documents that were sent via e-mail: they could not access these documents on the PDA. They therefore used e-mail on the PDA differently to the way they managed e-mails using the desktop or laptop computer. For example, the PDA would be used to deal with short e-mails, replying to messages that required simple "yes" or "no" answers or deleting e-mails that were no longer needed.

7.6.3 Note-taking

A small number of participants used the PDA extensively to take notes. Others attempted to use the PDA for this purpose but found the data input methods to be too slow and difficult. For those who persevered, however, the PDA provided a valuable way of creating and storing notes electronically. Participants who used the PDA as a note-taking tool were able to keep track of both personal and work-related information on the one device: "It is very useful for short notes to myself [...] So I'm using this like a little notebook all the time ... It is things I have to do but I don't put them down as to-do lists, I just have lots of these little notes ... It's everything from shopping lists as I'm wandering about thinking about what I need to buy when I go home tonight, to URLs to remember [and] passwords, so I'll take passwords with me when I'm logging on at other people's computers."

Many people kept the PDA either in their shirt pocket or in their handbag, so it was always available when they wanted to record or access information. Furthermore, unlike desktop or laptop computers which might, on occasions, be used by colleagues or family members, the PDA typically always remained in the hands of its owner. In this sense, as described above, some participants became quite dependent on the PDA as an information management tool. When used for this purpose, the PDA transcended work–home boundaries and became a truly personal tool.

Again, a further benefit of the PDA as a note-taking tool was that it could be used in conjunction with the desktop computer. Typically, notes recorded on the PDA were transient notes. They would be transferred to the desktop computer to be incorporated into another document or e-mail, or to be filed according to the user's information management system. This reduced the number of actions involved in recording and organising meeting notes: "I've scribbled notes in a meeting [on the PDA] and then come back to make notes from the meeting and of course they're now in electronic format as the outline for what becomes the word document that I just copy and move over, instead of again having a bit of paper that I work from."

That is, previously there had been a duplication of actions: notes scribbled on bits of paper then had to be typed up on the computer. The PDA simplified this process by enabling notes to be recorded in electronic format, which could then be uploaded onto the desktop computer when the two machines were synchronised. This facility was used extensively by one participant in particular who made substantial use of her PDA as a note-taking tool in combination with the foldout keyboard. This participant used the PDA and keyboard to take notes in seminars and conferences. Her ability to touch-type meant that she was able to record notes more efficiently than she would have done otherwise. For her, typing notes on a keyboard was an operationalised process and this freed up resources that could then be directed towards the action of listening to the speaker.

135

However, other participants found it difficult to utilise operationalised processes when taking notes on the PDA, particularly when using methods such as the handwriting recognition system or tapping out letters on the PDA's onscreen keyboard. This can be illustrated by an incident observed in which one participant attempted to use Graffiti on his Handspring PDA. As he had not yet learnt the specific Graffiti characters, the PDA did not recognise several of the letters he attempted. For example, he tried to write a lower case "a" whereas Graffiti only recognises this letter when it is written in upper case, as an inverted "v" (i.e. ∧). He also attempted to write a lower case "t", taking the stylus off the screen to cross the "t". The PDA interpreted this action as writing two separate letters. Exasperated, this person soon gave up writing anything on the PDA.

This incident is an example of what Bodker (1991) called a breakdown, caused by applying familiar operations to a new tool. What this participant did not realise is that he had to learn new ways of writing; that is, develop new operations, in order to make use of the Graffiti system. This incident can be contrasted with the experience of another participant who went to great lengths to learn the Graffiti alphabet: he did the onscreen tutorial included on the PDA and spent time playing a game designed to teach users Graffiti. Through this process, he developed the operations necessary to use Graffiti effectively and consequently said he found Graffiti very easy to use.

The PDA, then, initiated a number of changes in users' activity systems. How the PDA changed activities, though, was clearly dependent on the way the PDA was appropriated. In other words, the process of tool appropriation is closely related to the process by which activities are shaped. Together, these processes represent the two sides of the "reciprocal shaping" of technologies and society; this idea will be elaborated further below.

7.6.4 Tool Appropriation and the Shaping of Activities: A Two-way Process

The examples of PDA use in the two case studies described here (particularly the IET study) reveal that tool appropriation and tool mediation combine to create a two-way process. That is, while new technologies have an effect on the social setting into which they are introduced, features of the social setting also shape how new tools are used. This is illustrated in Figure 7.3.

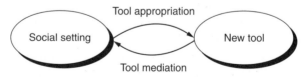

Figure 7.3 The two-way process of tool appropriation and tool mediation.

The idea that a two-way process occurs when new technologies are integrated into activities is not new. However, much of the literature on the social shaping of technologies is concerned with quite substantive issues. For instance, the essays included in MacKenzie and Wajcman's (1985) collection looked at how gender and socioeconomic issues have been embodied in technology designs. Elsewhere, researchers have examined how social factors affect technology use in third world countries (e.g. Akrich, 1992; Suchman, 2002) and how new technologies become domesticated at a national level (e.g. Brosveet and Sorensen, 2000).

The research presented here, in contrast, describes particular instances of PDA use. Departing from the more sociological research that occupies much of this field, the research did not examine the wider societal and cultural influences over the way PDAs have been adopted and used across broad social groups. Rather, the focus on specific case study settings and the use of interviews as the main source of data enabled a closer look at how the use of PDAs was socially shaped in individual cases. Furthermore, activity theory provided a way of identifying and describing this process of reciprocal shaping. Applying activity theory to the data helped to clarify the concepts of tool appropriation and tool mediation, emphasising that these dual concepts are closely related. In particular, the ASTAM framework was useful as a way of identifying the multiple interacting elements at play in the activity of appropriating a new tool. Although this was a departure from the more established uses of activity theory, analysing tool appropriation as an activity in itself was helpful for emphasising that no two users are exactly alike; each participant in this research used the PDA in a unique social setting. Tool appropriation is an individual activity as well as being socially situated. Recognition of this could help other evaluation studies examine why new technologies are often not used as expected.

While the idea of the reciprocal shaping of technologies and society is widely recognised in the literature, it is also often overlooked by research evaluating the use of technologies in learning and workplace settings (for instance, educational technology, HCI and information systems research). It is important for evaluation research to recognise the two-way process of tool appropriation and tool mediation as such an approach emphasises the broader picture of technology use. It is not enough to assess new technologies in terms of whether they add value, or in terms of how easy they are to use. The factors that make a device easy or difficult to use and the issues that contribute to how well the new tool is accepted and integrated into user's activities, will depend greatly on the context in which it is used, which varies from user to user. This was clearly demonstrated in the case studies described here, in which participants varied greatly in the way they responded to and used the PDA.

In order to understand how a new tool can be used effectively, then, and how that tool changes the activities it is used to support, it is important to look at the whole picture. This involves examining how new tools become integrated into the users' activities and evaluating the use of new tools from the perspective of the users themselves. Furthermore, researchers and technology providers need to appreciate that users will appropriate new technologies in their own unique ways. Therefore, use of new technologies cannot be prescribed and evaluations should not be based on predefined notions of how particular tools should be used. There is no "right" or "wrong" way to use a personal mobile technology, such as the PDA.

References

Akrich M (1992) The de-scription of technical objects. In: Bijker WE, Law J (eds). *Shaping Technology/Building Society*. The MIT Press, Cambridge, Massachusetts; pp. 205–224.

Bodker S (1991) *Through the Interface: A Human Activity Approach to User Interface Design*. Lawrence Erlbaum, Hillsdale, NJ.

Bodker S (1996) Applying activity theory to video analysis. In: Nardi BA (ed.). *Context and Consciousness: Activity Theory and Human–Computer Interaction*. The MIT Press, Cambridge, MA, London; pp. 147–174.

Brosveet J, Sorensen KH (2000) Fishing for fun and profit? National domestication of multimedia: the case of Norway. *The Information Society*, **16**: 263–276.

Carroll JM, Kellogg WA, Rosson MB (1991) The task-artifact cycle. In: Carroll JM (ed.). *Designing Interaction: Psychology at the Human–Computer Interface*. Cambridge University Press, Cambridge; pp. 74–102.

Carroll J, Howard S, Peck J, Murphy J (2002a) "No" to a free mobile: when adoption is not enough. In: Wenn A, McGrath M, Burstein F (eds). *Proceedings of the Thirteenth Australasian Conference on Information Systems (ACIS)*. Victoria University, Melbourne, Australia; pp. 899–910.

Carroll J, Howard S, Vetere F, Peck J, Murphy J (2002b) Just what do the youth of today want? Technology appropriation by young people. In: Sprague RH (ed.). *Proceedings of the 35th Annual Hawaii International Conference on System Sciences*. IEEE Computer Society, Maui, Hawaii.

Churchill EF, Wakeford N (2002) Framing mobile collaborations and mobile technologies. In: Brown B, Green N, Harper R (eds). *Wireless World: Social and Interactional Aspects of the Mobile Age*. Springer, London; pp. 154–179.

Engestrom Y (1987) *Learning by Expanding: An Activity–Theoretical Approach to Developmental Research*. Orienta-Konsultit, Helsinki.

Engestrom Y (1993) Developmental studies of work as a testbench of activity theory: the case of primary care medical practice. In: Chaiklin S, Lave J (eds). *Understanding Practice: Perspectives on Activity and Context*. Cambridge University Press, Cambridge; pp. 64–103.

Geisler C (2003) When management becomes personal: an activity–theoretic analysis of palm technologies. In: Bazerman C, Russell DR (eds). *Writing Selves/Writing Societies: Research from Activity Perspectives* [e-book]. http://wac.colostate.edu/books/selves_societies, Fort Collins, Colorado; pp. 125–158.

Halloran J, Rogers Y, Scaife M (2002) Taking the "no" out of lotus notes: activity theory, groupware, and student groupwork. In: Stahl G (ed.). *Computer Support for Collaborative Learning: Foundations for a CSCL Community – Proceedings of CSCL 2002* (Boulder, Colorado, USA 7–11 January). Lawrence Erlbaum Associates, New Jersey; pp. 169–178.

Kaptelinin V (1996) Activity theory: implications for human–computer interaction. In: Nardi BA (ed.). *Context and Consciousness: Activity Theory and Human–Computer Interaction*. The MIT Press, Cambridge, MA, London; pp. 103–116.

Kuutti K (1996) Activity theory as a potential framework for human–computer interaction research. In: Nardi BA (ed.). *Context and Consciousness: Activity Theory and Human–Computer Interaction*. The MIT Press, Cambridge, MA, London; pp.17–44.

Leont'ev AN (1978) *Activity, Consciousness, and Personality*. Prentice-Hall, Englewood Cliffs.

MacKenzie D, Wajcman J (eds) (1985) *The Social Shaping of Technology: How the Refrigerator Got its Hum*. Open University Press, Milton Keynes.

Nardi BA (1996) Activity theory and human–computer interaction. In: Nardi BA (ed.). *Context and Consciousness: Activity Theory and Human–Computer Interaction*. The MIT Press, Cambridge, MA, London; pp. 7–16.

Sears A, Arora R (2002) Data entry for mobile devices: an empirical comparison of novice performance with jot and graffiti. *Interacting with Computers*, **14**: 413–433.

Suchman L (2002) Practice-based design of information systems: notes from the hyperdeveloped world. *The Information Society*, **18**: 139–144.

Turner P, Turner S (2001) A web of contradictions. *Interacting with Computers*, **14**: 1–14.

Vygotsky LS (1978) *Mind in Society: The Development of Higher Psychological Processes*. Harvard University Press, Cambridge, MA, London.

Waycott J (2002) Reading with new tools: an evaluation of personal digital assistants as tools for reading course materials. *ALT-J*, **10**: 38–50.

Waycott J, Kukulska-Hulme A (2003) Students' experiences with PDAs for reading course materials. *Personal and Ubiquitous Computing*, **7**: 30–43.

Wertsch JV (1998) *Mind as Action*. Oxford University Press, New York, Oxford.

Part 3
How to Study the Future

8

Different Directions in the Mobile Internet: Analysing Mobile Internet Services in Japan and Europe

Richard Tee

8.1 Introduction: The Mobile Revolution in Retrospect – *Contradictio in Terminis?*

Is it possible to reflect back on a revolution we are currently in the midst of? This chapter will present one attempt to do so. It will in particular look at developments in the mobile Internet in Japan and Europe that have taken place from the late 1990s. In the process this chapter hopes to generate lessons that are valuable for industry, academia and policy makers. In comparison with Europe, Japan has experienced high levels of take up of mobile Internet services. This chapter will attempt to summarise various explanations and will discuss their merits. The factors brought forward with regard to the success of the mobile Internet in Japan usually focus on technical, cultural or market structural differences between Europe and Japan. Market structure appears to be a key factor. The mobile Internet in Europe is more fragmented. It begins with the success of SMS (Short Messaging Service) and the subsequent failing of WAP (Wireless Application Protocol). It is proposed to interpret these developments by distinguishing between three different types of approaches: protocol, service and platform based. Examples of protocol-based initiatives are SMS, WAP and MMS (Multimedia Messaging Service). Next there are service-based initiatives, which use the Japanese business model and approach; i-mode Europe and Vodafone Live are two notable examples. Thirdly there are platform-based approaches such as the Microsoft Smartphone and the Symbian operating system.

One of the main premises of this chapter is that technology is shaped by social factors, a standpoint derived from the conceptual framework called the Social Construction of Technology (SCOT). One of the basic assumptions within the SCOT approach is that technological developments cannot be addressed satisfactorily by focusing solely on technology. Rather, the breadth of explanation should be expanded to include social, political, economic, cultural and technological factors, which together shape the ways technological innovations evolve. The main objective in this chapter will be to identify the underlying causes that have shaped and are shaping the development of mobile Internet services in Europe and Japan.

In order to address the research question, both primary and secondary research have been conducted. The secondary research has consisted of a literature study from the typical variety of sources, including journal articles, books, periodicals and web publications. Primary research consisted of more than 25 interviews with both users and producers of mobile Internet services. "Producers" should be interpreted widely and does not refer strictly to technical developers of mobile Internet services. Therefore, this category also includes researchers, financial analysts, consultants, venture capitalists, etc. Other people that were interviewed include telecom operators, content providers and software developers.

Defining a mobile Internet service such as i-mode in conventional terms has proven to be surprisingly difficult. In technical terms one could say i-mode provides the mobile phone with a network interface. Besides a term such as network interface, there have been other ways to label this shift from voice to data usage. Sometimes it is referred to as mobile data services, whereas others have referred to it as the wireless Internet. The most prevalent term used at the moment is "mobile Internet". The term itself is not without problems. One of the main problems is that the term conveys that what is referred to is the Internet, but then mobile. It is argued that this is a misleading representation when considering the ways people use mobile Internet devices. For example, the most popular uses of i-mode are sending messages, downloading ringtones and pictures, playing games, reading horoscopes, looking up news and weather information and dating services (Funk, 2001). Despite these differences from the fixed line Internet, for the sake of convention and clarity "mobile Internet" is used as the main descriptor.

Note the difference between the terms "mobile Internet" and "mobile Internet services". The convention is that services such as i-mode, Vodafone Live and others are – quite logically – referred to as mobile Internet services. WAP, MMS or the Microsoft Smartphone are not services, but protocols and a device respectively. Referring to these then requires some creativity with words; in this context they are referred to as instantiations, conceptualisations or implementations of the mobile Internet. The shorthand "mobile Internet" or "the mobile Internet" is usually used as an umbrella term, to refer both to mobile Internet services as well as the protocols and devices that are associated with them.

8.2 Conceptual Framework: Social Construction of Technology

The theoretical perspective of this chapter is informed by the SCOT approach. Central to this approach is the idea that technological innovations, of which the mobile Internet is an example, are socially constructed. SCOT can be regarded as a response to a perspective on technology and technological innovation referred to as technological determinism. It consists of two parts. The first is that technological innovations develop outside of society, uninfluenced by social, economic or political forces. The second part is that technological developments cause or determine social changes, in a fixed and linear fashion (see e.g. Williams and Edge, 1996). Critics of technological determinism, such as SCOT theorists, reject these claims. One of the shortcomings of technological determinism is that it is based on an overtly simplistic, reductive way of thinking. Technology is represented as the single independent variable, which then unidirectly impacts on society. This representation leaves out many crucial aspects of the way technology evolves.

A classic example from the SCOT literature is the development of the low-wheeled bicycle. Bijker (1990) demonstrates how the bicycle can be viewed as a social construction. The first bicycles were the high-wheeled versions, used mainly by well-off young men, who demonstrated their skills in parks and bicycle contests. The bicycle was regarded as a vehicle for speeding, rather than a method of transportion. Learning to ride these high-wheeled bicycles proved to be – literally – an obstacle for older people as well as youngsters. Furthermore the Ordinary, as the high-wheeled bicycle was called, was not considered suitable for women due to the dressing customs and morals of the time. For these groups of non-users, the bicycle was regarded as an unsafe machine. Further innovations brought along the air tyre. One of the advantages of the air tyre was that it worked as an anti-vibration device and therefore made riding a bike more comfortable. However, for the main users of the time the bicycle was regarded as machine for "speeding" and thus they had no interest in the air tyre. But the air tyre, which was used on the low-wheeled bicycles, often scorned by high wheelers, had another advantage: it permitted higher speeds. Consequently, the air tyre was transformed from anti-vibration device to speed device and thus the high-speed tyre was born. Of course, the high speed tyre did appeal to the existing bicycle users. Because it catered to both groups, the design of the bicycle stabilised to the low-wheeled type as we still know it today.

Central to SCOT thinking is that people have choices regarding the form and use of a particular technology. Main concepts in the SCOT theory are relevant social groups, technological frame, interpretive flexibility and closure. Taking the bicycle example, the relevant social groups are for example engineers, bicycle users and non-users, people opposing bicycles and advertisers. It is important to make a distinction between the users (initially young, well-off males) and non-users, since their interests were so different. This relates to the concept of interpretative flexibility. For the first group of users, the bicycle was a "macho machine", a speed demon. For them the technology "worked". For others it was an unsafe machine, risky to learn and unsafe to ride. For them the technology did not work. The technological frame provides the framework in which relevant social groups operate. It is similar to the concept of paradigm as conceived by Kuhn and provides "the goals, the thoughts, the tools for action. They guide thinking and action." (Bijker *et al.*, 1987).

8.3 Case 1: The Mobile Internet in Japan

8.3.1 Background to the i-mode Project

The i-mode started out as the "DoCoMo gateway project" (Matsunaga, 2002; Natsuno, 2003). This project started in 1997, headed by Keiichi Enoki, who was appointed as leader of the project by DoCoMo CEO Kouji Ohboshi. At the time NTT DoCoMo, a play on the Japanese phrase "doko demo", meaning "everywhere", was not regarded as a particularly promising department by parent company NTT (Beck and Wade, 2002: 79; Ratliff, 2000). Mobile phones at that time were not handheld phones but car phones and statistically, mobile telephony was a micro-niche, with a penetration of 1% (Beck and Wade, 2002). Enoki, Matsunaga and Natsuno are usually identified as the key persons in the development of i-mode. Matsunaga (2002) describing how she is trying to persuade Natsuno to join DoCoMo provides insight into the way i-mode (at that time unnamed) was thought of: "You know those LCD displays they have on mobile phones? Well, we're thinking of using them to distribute info from the Internet, but you know how hopeless I am with technology. Will you help me out?"

It could be argued that the sentences above provide i-mode's technological frame. Its proposition was very simple: distribute information from the Internet using LCD displays from mobile phones.

The initial response to i-mode was not overwhelming. As reported by Matsunaga (2002) and elsewhere, for the original press conference announcing the release of i-mode, only seven reporters showed up. While the launch was delayed anyway, it was decided to host a second press conference, this time using a different strategy. Japanese TV celebrity Ryoko Hirosue was chosen to star in the i-mode commercials, preceding the actual launch of the service. This time the i-mode team did manage to get the press interested; the second introductory launch on 25 January 1999 attracted over 500 reporters. Nevertheless, the uptake of the service started slowly. The service was introduced 22 February 1999, and a month after the release, there were only 20,000 subscribers. However, the growth thereafter turned out to be explosive: by the end of June 1999 there were 500,000 subscribers. In early August i-mode

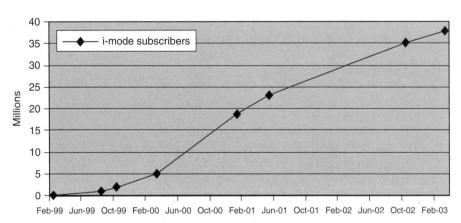

Figure 8.1   The i-mode subscriber growth in Japan.
Source: DoCoMo; Funk, 2001.

reached 1 million users. By March the following year there were 5 million users; August 2000 marked 10 million subscribers. As of August 2004, the number of subscribers has grown to approximately 42.2 million users (Mobile Media Japan, 2004) (Figure 8.1).

Next to i-mode, there are two competing services, J-Sky and KDDI's EZweb, both of which have approximately 10 million users. Although they are not as large as DoCoMo, both carriers have offered services that have influenced the development of mobile Internet services. J-Phone has done this most notably by being the first to offer 2.5G camera phones and has attracted more users with its associated photo messaging service Shamail. KDDI has received praise for their 3G upgrade path with which they managed to attract about 7 million users, more than 20 times as many as DoCoMo's 3G service called FOMA.

8.3.2 i-mode Explained: Different Explanations

An important question is why mobile Internet services managed to reach such high levels of diffusion in Japan. Partly as a response to the lack of uptake of WAP services in Europe, many articles have been written trying to establish the key factors for this success. Some analysts provide technical arguments, where the packet-based network of DoCoMo is regarded a key success factor (Ducont, 2004). Others maintain it is a cultural issue and point at the presumed fondness of Japanese consumers for new technologies (Allnetdevices, 2001). Another often heard statement is the supposed fact that Internet connections were significantly lower at the time of introduction of i-mode compared to other countries where mobile Internet services where introduced, thus facilitating its uptake

147

(Allnetdevices, 2001). Three basic questions are usually addressed in these "success of i-mode" articles (see e.g. Chiu *et al.*, 2001):

- why it succeeded;
- whether this success is limited to Japan;
- if not, how this success can be exported to other countries.

This part will discuss each of these arguments and will assess the merits and limitations of them. The different types of arguments can be divided into three main categories, technical (network type and markup language), cultural (fondness for gadgets and commuting habits) and market structure (low PC Internet diffusion, micropayment system and revenue sharing model, originator mobile phone specifications).

One of the most frequently heard arguments as to why mobile Internet services succeeded in Japan was due to the availability of a packet-switched data network (see e.g. Allnetdevices, 2001). A network, such as a mobile telephony network, can either be packet-switched or circuit-switched. A circuit-switched network ensures there is a dedicated connection between two points. In a packet-based network packets of data are sent between two points but there is no continuous connection required. Traditional telephony networks have usually been circuit-switched since it is crucial that there is a continuous and reliable connection. The regular Internet is based on packet-switching. Relevant in this context is that in a packet-switched network, there is no dial-up time when connecting to a network, which is present in a circuit-switched network. Furthermore, it is much easier to bill users for the amount of data exchanged rather than for the time spent on the network. The latter billing scheme will, given the structure of the network, be more expensive to the end-user than a time-based solution. Because i-mode was run over a packet-switched network, it has frequently been argued that the network type has been crucial for the success of mobile Internet services. Given the success of J-Phone's mobile Internet service which was initially using a circuit-switched network, but used packet-like billing, it could be argued that it is packet billing that was the key issue i-mode for the success of mobile Internet services. Note that the packet/circuit argument is usually brought forward not only with regards to i-mode. The introduction of WAP services in Europe and elsewhere is usually the starting point of the packet/circuit discussion, while WAP services were mostly introduced over a circuit switched network and generally did not take off as expected.

Aside from the packet/circuit argument, there exists another technical argument as to why i-mode took off. The i-mode uses cHTML as a markup language, rather than, for example, WAP services which use WML. The argument for or against cHTML as key enabler follows a similar logic as the packet/circuit argument. Proponents of the cHTML as enabler perspective, point at the apparent failure of WAP services, which by default used WML as markup language. More informed opponents point at KDDI's EZweb which uses WML and has attracted 11 million users as well as a developed content base. This content base is the key issue with regard to markup language. cHTML proponents maintain that it was the compatibility with HTML, the standard markup language used on the fixed line Internet, that ensured content providers were willing to create i-mode services. Opponents state that WAP 2.0 is compatible with cHTML and it is thus irrelevant what markup language is used. Like in the packet/circuit argument a distinction has to be made between i-mode and mobile services in general. Was cHTML a requirement for i-mode to succeed? Ultimately this question could be answered positively, for the simple reason that it lowered the barrier for content providers to take part in the i-mode scheme. As can be read in Matsunaga's account of the creation of i-mode, initially it was difficult for DoCoMo to convince content providers to create services (Matsunaga, 2000). Even though the differences between WAP and cHTML are not dramatic (and upgrades to the WAP specification have made them compatible), it has been argued that the difficulty of having to use another channel for a new service does create a psychological threshold. Since DoCoMo chose cHTML, content providers did not have to learn an entirely different language and could also use the existing infrastructure they had been using for their fixed line services. Thus, it can be argued that the choice against WML (even though DoCoMo used to be a boardmember of the WAP forum) and for cHTML was indeed a requirement for i-mode to develop the way it did. This is different from the issue of whether it is a requirement for mobile services in general. Both J-Phone and KDDI show it is possible to introduce a successful service using another language besides cHTML. However, their models were only viable because of the initial success of i-mode. As with the circuit/packet issue, the historical trajectory is crucial. Packet billing was indeed initially a very important factor, a key factor, for i-mode. After its success services could also be introduced from a different type of network. Essentially, however, the service is the same, in terms of the business model, content and marketing. The ground work done by DoCoMo with its i-mode service lowered the barriers for other service providers, who could thereafter introduce their respective services using slightly different standards. Thus, it is incorrect to argue a WAP-based service such as Vodafone Live cannot succeed. However, initially cHTML was indeed a key factor. After DoCoMo demonstrated the viability of its business model other service providers could succeed using alternative standards.

A cultural stereotype has it that the Japanese have a fondness for new technologies that is unique in the world. As usual with common sense statements, there is at least some truth to be found in this claim, which is also backed up by experts on Japanese society (see Tee, 2002). The downside of most common sense statements is lack of subtlety. It is not difficult to find examples of technologies that have reached some degree of popularity in Japan and have not taken off elsewhere. Three well-known ones are the Tamagotchi virtual pet, its pumped up cousin (in terms of both price and capabilities) the Sony Aibo dog; and last but not least pre-warmed electronic toilet seats with built-in flush sounds to avoid any neighbouring bathroom users from hearing embarrassing sounds. Granted, the first two examples are strictly speaking not entirely Japanese. Tamagotchi was apparently popular outside Japan as well, but to nowhere near the degree it once was in Japan. The same goes for the Sony Aibo pet dog, even though there are also fanatic collectors in other countries. The point is that there are innovations which are more popular in Japan than elsewhere. The question is of course whether i-mode, or mobile services in general, belong to that category or to the group of Japanese innovations that did succeed outside of the country like the Walkman or Gameboy. Given the popularity of downloadable ringtones, screensavers and SMS in Europe, it seems difficult to argue that mobile services as they exist in Japan are confined to its borders.

It has been claimed that Japanese users were more receptive to mobile services because of commuting habits, where public transport is used more than in other countries. Validating this claim requires a comparison of the modes of transportation for both Europe and Japan. One indicator is car ownership. Data from the US Department of Transportation shows that in Japan, the cars per person ratio is 0.327 compared to 0.424 for France, 0.465 for Germany, 0.363 for the UK and 0.409 for Sweden (Federal Highway Administration, 1993). Based on these findings it could be argued that the relatively low car ownership rate might have played a role in the uptake of mobile Internet services. Of course, car ownership must be distinguished from car usage. Therefore, it is helpful to look at commuting habits. Data from the respective departments and ministries of transportation give more insight into this matter. Table 8.1 shows the average numbers for daily public (bus and railway) and private (car) transport for Europe, the US and Japan respectively.

The numbers clearly show how Japan leads in terms of public transportation usage rates (36.4%) against 16.0% for the European Union and a mere 3.9% for the US. Given these figures, it can be concluded that commuting habits have played a facilitating role in the general uptake of mobile services in Japan.

Apart from technical network or cultural reasons there are other factors relating to market structure that might have played a role in the uptake of mobile services. One claim is the rate of fixed Internet connections and the

Table 8.1 Public and private transport usage in Europe, Japan and the US in 1998

	Europe	Japan	US
Private transport usage (%)	79.2	58.2	85.6
Public transport (%)	16.0	36.4	3.9

Source: The European Union, US Department of Transportation and Japan Ministry of Transport. The percentages do not add up to 100% because of the exclusion of air transport. http://www.publicpurpose.com/tfb-cheujpus98-pkm.htm

apparent lack of these during the time of introduction of i-mode. Others point at the micropayment and billing system provided by NTT DoCoMo. Another important issue is control of phone specifications.

During the introduction of i-mode, fixed line Internet penetration in Japan was apparently lower than other countries. This might have facilitated people's willingness to try i-mode and thus increase the rate of diffusion. In this case it will be useful to look at data for fixed line Internet penetration, around the time that i-mode was introduced (early 1999). Figure 8.2 shows that Japan's fixed line Internet penetration was in fact not significantly lower than other countries with penetrations that are considered to be "high", such as the UK, Germany or South Korea. Supposed lack of fixed line computer and Internet usage has proved to be one of the most common myths with regards to the success of Japanese mobile Internet services (Allnetdevices, 2001).

Even if the penetration of the fixed line Internet had been significantly lower, it is still arguable that this did not contribute to the uptake of mobile Internet services for two reasons. The first is the premise behind the low-fixed-Internet logic, which is that fixed and mobile Internet are substitutes. For instance, the major application on the mobile Internet, just like on the fixed line Internet is mailing and messaging. However, apart from that, ringtones and screensavers are most popular, which are certainly different applications from the fixed Internet. Rather than substitutes, it is arguable that fixed and mobile Internet access are complementary services. This view is supported by the notion that use of the fixed line Internet in Japan has increased very rapidly since the introduction of mobile services. Thus the logic is reversed; it is the diffusion of mobile services that has actually spurred the uptake of fixed line Internet. Another example supporting the complementarity view is the initial interest present during the introduction of WAP services. Although both the US and Europe had high fixed line Internet penetration, interest in mobile services was very high, at least

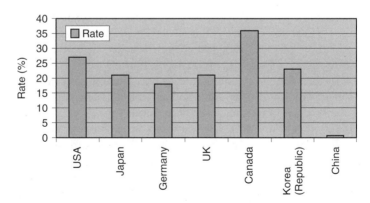

Figure 8.2   Internet penetration rate, 1999.
Source: ITU (2000) from Pikula, 2001.

151

as measured by the amount of media attention. The fact that the usage remained dramatically low might have had more to do with the actual implementation of the service rather than lack of interest. It appears that there is no causal relation between low fixed line Internet penetration and uptake of mobile Internet services. Rather the two are viewed as complementary services.

The revenue sharing and micropayment system NTT DoCoMo introduced with i-mode has been widely praised. It has also been identified as one of the key enablers behind the success of i-mode. Lately there has been some discussion about the micropayment system itself, such as the revenue sharing distribution between network operator and content provider. However, the importance this model has had on the availability of content remains unchallenged. Unanimously it is agreed that this move has been essential for the success of i-mode.

DoCoMo hired the renowned consultancy agency McKinsey to help with the i-mode business strategy. Matsunaga (2002) chronicles the internal battles between the McKinsey consultants and the regular i-mode team. Some suggestions had been made by the consultants, which would have greatly affected the available content. For instance, McKinsey suggested that DoCoMo charge firms for a position on the portal (and let the content providers themselves be responsible for fee collection) and in other cases that DoCoMo could also purchase content it absolutely required (Matsunaga, 2002: 97, 98; Funk, 2001: 23). Another issue was the setting of a maximum fee for content providers. Both McKinsey as well as Natsuno agreed a 1,000 yen (about £5) per month fee would still be reasonable. Matsunaga insisted it should be far lower than that. After long discussion, Matsunaga finally convinced both McKinsey as well as other team members that 300 yen (about £1.50) should be the maximum, which she based on the average price for a magazine as well as her rounding off theory: "Five or more and you round up, four and below and you round down" (Matsunaga, 2002).

In Japan, it is the carrier who largely coordinates the technical specifications of the phone. This situation is rather unusual. For instance in Europe or the US the handset makers themselves decide the characteristics of next year's models, in terms of design, size, functionality, type of display, etc. With the introduction of i-mode, it was DoCoMo who specified to the handset makers what their i-mode phone should be like. These specifications go as far as the weight, volume, display resolution and markup language of the phone. Thus, when DoCoMo released their first i-mode capable phone, they were certain the content of the service was entirely compatible with the phone.

In summary, how can the success of i-mode and other mobile Internet services be explained? Why did it "work" for Japanese users? First of all, it is very important to distinguish between i-mode, the leading service, and mobile Internet services in general. In terms of network factors, packet-switching and the choice of cHTML are identified as key factors for the success of i-mode. Packet billing appears to be a requirement for mobile services in general. The choice of markup language does not seem to be a significant indicator for mobile Internet services introduced after i-mode, given the comparable success rates of EZweb and J-Sky. As for cultural factors that might have facilitated the uptake of mobile services, an inclination towards electronic gadgets is in this case largely irrelevant. Some factors related to differences in market structure have played a major role. The claim that fixed line Internet diffusion was significantly lower in Japan proves to be unfounded. What remains are two key issues which have played a major role in facilitating the uptake of mobile services. The first is the idea of revenue sharing, combined with the creation of a micro-payment system and a portal. Second is the level of coordination that Japanese telecom operators, most notably NTT DoCoMo, can exercise over other relevant parties.

8.4 Case 2: Mobile Internet in Europe

8.4.1 Successful SMS, Failing WAP

SMS so the story goes, was added as an afterthought to the Global System for Mobile Communication (GSM) specification (Shahin *et al.*, 2002). Hence the immense success of SMS came as a complete surprise, for telecom operators, handset makers as well as content providers. In the case of SMS most of the revenue comes from peer-to-peer messaging as well as additional traffic generated by third-party service providers. These third-party applications range from downloads such as ringtones and screensavers to horoscopes, game shows and railway information. Operators hoped and expected a similar scenario would emerge with the introduction of the protocol WAP. WAP and its associated information delivery services failed, for many reasons, ranging from technical, to industry-structure-wise to business-wise. Several other standards were proposed as the follow-up to SMS. There were open standards such as EMS (Enhanced Messaging Service) and MMS. Both EMS and MMS have been developed by the 3GPP (3G Partnership Program) standards body. Nokia developed their proprietary protocol called Smart Messaging.

8.4.2 Interpreting Story 1: Arguments for Success and Failure

As mentioned earlier, both the success of SMS as well as the failure of WAP were unexpected. The important question is of course why SMS worked for a large group of users, while WAP did not.

Several reasons can be brought forward to account for the success of SMS. First of all, SMS was a standard feature on virtually all GSM phones. Furthermore, SMS is essentially a peer-to-peer service and it is not dependent on any third-party content. Even without the number of complementing services that are offered, SMS can still be functional as a pure messaging service. Another often mentioned reason is the price of sending an SMS, compared to the price of making a call. It is difficult to make a straightforward comparison between calling and sending an SMS, since calling charges and data charges can vary dramatically, depending on factors such as calling plan, type of number, location, time of calling, etc. As a general rule however, the price of sending a message will in most cases be cheaper than calling. Complementing and supporting the popularity of text messaging, several devices and technologies have been introduced to allow faster and easier text composing. This includes what is called predictive text input and physical devices such as add-on keyboards, as well as the introduction phones that featured a (phone-sized) full size keyboard.

Several reasons have been forwarded to explain why WAP services did not take off as expected. A report released by the International Telecommunications Union (ITU, 2002) cites extended waiting periods for downloading, ineffective billing models, lack of content availability in WML and inappropriate (monochrome) interface for viewing Web content. The origin of WAP has been chronicled thoroughly by Sigurdson (2001). WAP was conceived and initially developed by two organisations: Ericsson and Unwired Planet (now called Openwave), a US company specialising in the development of minibrowsers. A crucial issue for the viability of WAP was to convince other key players to join the WAP Forum. Apart from Ericsson, the established members had to persuade the major makers, Motorola and, most importantly, Nokia, to join. Motorola decided to join. Convincing Nokia proved to be more difficult. During the same period when EMS was being developed Nokia was working on their Smart Messaging protocol and was in doubt whether or not to join the open standard WAP. Either they could join what was expected to become the *de facto* standard for the mobile Internet, or stay out and work on their own proprietary standard. The inclusion of Nokia was unclear until the day before the first press conference. For that reason two press releases had been prepared, one with Nokia having joined, the other one without Nokia. Finally, Nokia decided to join. Nevertheless they continued with Smart Messaging, which they offered jointly with their WAP phones. After Nokia had joined, there were other big players to persuade, most notably Microsoft and NTT DoCoMo. Both initially did not join the WAP Forum, for different reasons. Microsoft was involved with their own version of an operating system for mobile devices, Pocket PC. Pocket PC is used mainly on Personal Digital Assistants (PDAs) now and Microsoft did not see the benefit of joining a standard limited to mobile phones only (Sigurdson, 2001). Of course, currently Microsoft has developed their own OS for mobile phones, released via their Smartphones initiative, more on which later. NTT DoCoMo was working on their own, at that time unnamed, mobile Internet service. Both Microsoft and DoCoMo decided to join later, and DoCoMo even became a board member (Sigurdson,

2001). WAP was plagued by difficulties, some of which were mentioned earlier. In addition to these, there were usability issues, initial lack of handsets and the prices of these handsets (Sigurdson, 2001).

8.4.3 Mobile Services Continue: i-mode, Vodafone Live, MMS, 3G, Smartphone

After the disappointing uptake of WAP services, operators were looking for new channels to increase ARPU (average revenue per user). Trying to generate additional revenue by introducing non-voice services still seemed an obvious choice, given the successes obtained in Japan. The telecom operators could pursue several strategies. One was to wait for the deployment of 3G networks and hope for complementing "killer applications" to proliferate. However, the roll out of 3G networks took longer than expected. Whereas initial projections predicted that 3G networks would have become mainstream by 2003 (see e.g. Chiu *et al.*, 2001), by April 2003 only one actual 3G service had been introduced (from operator "3", owned by Hutchison).

Another option, which potentially enabled companies to increase ARPU and at the same time decrease churn, was to invest in services running on 2.5G networks such as GPRS (General Packet Radio Service). Here, three general types of approaches can be distinguished: protocol, service and platform based. Examples of protocol-based initiatives are SMS and WAP and also MMS. Service-based initiatives are for example i-mode and Vodafone Live. Symbian, as well as the Microsoft Smartphone are good examples of platform-based proposals. These labels are conceptual. In practice, protocols can overlap with services (such as the inclusion of MMS within Vodafone Live), services with platforms (e.g. the Series 60 based Nokia 7650 and Vodafone Live) and all other combinations.

MMS was introduced in a low key manner, compared to the WAP spectacle several years before. One of the expected "killer applications" of MMS is sending pictures taken with the handset's built-in camera. MMS has the advantage that it is an open standard, and is fast becoming a *de facto* standard on all newer handsets. One crucial advantage of MMS is that it is also supported by Nokia, something which was not the case with EMS for instance. Then again, using an industry supported standard does not automatically mean success, as witnessed by the roll out of WAP services. There are other advantages to MMS besides being an open standard. The acronym itself has been chosen to reflect MMS's link to SMS, since SMS is the most widely known of all mobile Internet standards. Besides being "acronymatically" linked to SMS, there is also real world compatibility. This compatibility is reached either via SMS or e-mail. Proponents of MMS argue this will enhance the interest in MMS. Another advantage for MMS is that it is not outwardly in competition with other services. Vodafone has embraced MMS in their Live service from the start. The i-mode did not initially but has decided it will support MMS in newer i-mode handsets, something that can be considered a small victory for MMS. Besides being supported by both Vodafone Live and i-mode, an MMS client is also integrated in the Microsoft Smartphone.

The second strategy was to employ the Japanese approach. This means creating a product, a service, rather than embracing a protocol, and releasing handsets that support the service as well as creating a platform that allows content providers to easily distribute (and charge) for their services.

The i-mode was first released in Germany in March 2002 by E-plus, a subsidiary of Dutch incumbent operator KPN. The Netherlands followed one month later. The i-mode team did a good job arranging publicity. All major Dutch newspapers and magazines wrote about i-mode and it was even mentioned on the 8 o'clock news, a privilege for an IT product only Windows 95 experienced before (Van der Molen, 2002). Despite the publicity, the initial uptake was fairly slow, which some people attributed to the fact that initially only one type of handset was available (though we should note that in Japan also initially just one handset was available). In December 2002, nine months after the release, i-mode reached 100,000 subscribers in Germany, the Netherlands and Belgium combined (Luna, 2002).

Some blame the rather slow uptake of i-mode on the fact that none of the major handset makers participated. The first model released was an NEC handset; the second, a Toshiba. Both handset makers had never released a phone before in Europe. The NEC sported the "clamshell" design, by far the most popular design in Japan. Europeans are said to prefer the "candybar" look, prevalent in all Nokia and Siemens handsets for instance. Thus, the Toshiba phone had the candybar look. Market leader Nokia has consistently refused to release an i-mode phone, despite attempts from KPN to persuade them to join (Volkskrant, 2002). Also, brand awareness works very differently in Europe. Handset awareness certainly does exist in Japan, but the emphasis lies on the service provider, not on the handset brand. In Europe the opposite is true, with disproportionate weight placed on the handset brand. Therefore, the phone has been and still is the major consideration for most users. And since up to this point none of the major handset suppliers has yet chosen to support i-mode, this is limiting the potential exposure for i-mode.

Vodafone Live was introduced by global operator Vodafone on October 2002 in six countries at once. It offered three different phones, all of which featured a built-in photo-camera. Generally speaking, the Vodafone Live approach was very similar to the i-mode model. It made use of a portal, a micropayment system and revenue sharing model, both official and unofficial sites and dedicated handsets. An important difference for content providers was that unofficial content providers were also allowed to make use of the micropayment system. The revenue sharing agreement between network operator and content provider, something which DoCoMo has emphasised from the beginning, has not been disclosed for Vodafone Live, though it is reported to be "along the lines of" the 91–99% division used for i-mode. The uptake of Vodafone Live was better than for i-mode in absolute numbers: it reached 1 million subscribers by March 2003 (Van Impe, 2003). However, the comparison is difficult to make while the market sizes and penetration rates are different. Whereas KPN initially did not succeed in persuading Nokia to release an i-mode phone, Vodafone did manage to have one Nokia in their Vodafone Live lineup. Apparently, Vodafone Live was more successful than i-mode in terms of handset sales in the first months after release. Ironically, the most popular handset was not the Nokia but the Sharp handset.

Besides i-mode and Vodafone Live, there have been other entries into the world of mobile Internet, two notable ones being the O2 XDA and the Orange SPV Smartphone. What they both have in common is that they are running on Microsoft operating systems. The XDA, part PDA, part phone, was never meant to be a mass market product and failed to reach high sale numbers. The Orange SPV (Sound, Pictures, Video) however was conceived as a consumer product. It was released in France first by Orange, manufactured by Taiwanese maker High Tech Computer (HTC). Originally, British maker Sendo had planned to produce the first Smartphone, but bailed out at the last minute, accusing Microsoft of sharing Sendo's trade secrets and proprietary technologies (Evers, 2004).

When it became clear to handset makers that Microsoft was planning to enter the market of mobile devices, vendors decided to bundle their strengths in an alliance called Symbian. Headed by Nokia, and joined by a number of other major handset makers, the goal of Symbian is to provide an open platform for smartphones. Rather than smartphones being equipped with a Microsoft OS, with Symbian handset makers can use Symbian as their base OS and offer the system according to the preferences of the handset maker.

An advantage of the Microsoft approach is that it is, arguably, the most literal translation of mobile Internet. Most of the major applications known from the traditional Internet (e-mail, web and messaging) are present in the Smartphone approach. More importantly, it also brings along two very popular services from the fixed line Internet, Hotmail and MSN Messenger. Microsoft depends mainly on two relevant social groups in their Smartphone offerings. The first is manufacturers who are willing to make Smartphone handsets. The second group are operators that are interested in releasing these phones. The second condition depends on the first. That is, in most cases it will depend on which handset maker makes the Smartphone. If it would be Nokia, all operators would probably be willing to offer the phone. If it is a made-to-order company such as HTC who makes the phone the situation might be different. Right now it has been Orange who has branded the phone as an Orange Smartphone, ignoring the maker in this case. Out of all the major makers, there has only been one maker, Samsung, which has announced the release of a Smartphone. It will be interesting to see whether this single operator branding will sustain with the release of the Samsung Smartphone.

8.5 Conclusion: Mobile Internet Services – a Work in Progress

Going back to the main question, when we compare Europe with Japan, what are the underlying causes that have shaped the development of the mobile Internet in such different ways? A key issue brought forward is market structure, where the role of the mobile operator plays a significant role. In Japan, the operator plays the key role in coordinating and – some would argue – thereby controlling the other partners, which includes control of content providers as well as handset makers. Mobile services in Japan begin and end with the mobile operator. The situation in Europe is quite different. There is no party that clearly dominates. It could be argued that handset makers have an instigating role with regards to the development of technical features. However, the handset maker neither has the interest nor the capacity to play the role the operator has done in Japan. This appears to create a tension between operator and handset maker, as demonstrated in negotiations going on between i-mode operators, Vodafone and handset makers such as Nokia.

The current movement that is taking place in the evolution of the mobile Internet demonstrates the concept of interpretive flexibility. Three types of initiatives – protocol, service and platform-based approaches – can be identified. The Smartphone, an example of a platform-based initiative, is based on a Microsoft operating system. Rather than a phone with Internet-like capabilities it can also be conceptualised as a small computer with phone capabilities. In different ways, Microsoft and other mobile operators are offering ways in which the mobile Internet can be implemented. Symbian is an alternative platform-based initiative, Series 60 being an important open UI for this. The i-mode and Vodafone Live are examples of a service-based approach to the mobile Internet. The i-mode originates from a phone company, made by traditional handset makers and is mostly equipped with proprietary software. The i-mode and its related cousins have been largely positioned outside of the traditional desktop PC. It can be thought of as a phone with additional Internet-like features, including features such as mobile Java. MMS is an open standard which can be integrated in any of the three ser-vices mentioned earlier, or be offered independently. While the Smartphone as well as i-mode type services are all proprietary platforms, MMS is not. Rather it is an open standards protocol that is supported by all handset makers, including the Smartphone as well as Vodafone Live handsets and i-mode models. Stabilisation has not occurred in the world of mobile Internet. Of all these different standards, protocols and operating systems, the key question will be to see which set of interpretations will prevail.

References

Allnetdevices (2001) http://www.allnetdevices.com/wireless/opinions/2001/10/01/exporting_i.html
Bijker WE, Hughes TP, Pinch T (eds.). (1987) *The Social Construction of Technological Systems*. New Directions in the Sociology and History of Technology. MIT Press, Cambridge, Massachusetts.

Bijker WE (1990) *The Social Construction of Technology*. Eijsden: Proefschrift Universiteit Twente.

Chiu PL, Guell M, Wanninger C (2001) Mobile Internet – For technique freaks or Mass Market? Infocom Research Programme, MA Thesis Lund University.

Computerwire (2003) HTC, T-Mobile to launch Orange-like MS smartphone. The Register. http://www.theregister.co.uk/2003/01/06/htc_tmobile_to_launch_orangelike/

Ducont (2004) http://www.ducont.com/Technology/FAQs/tech_imodefaqs.htm

Evers (2004) Sendo, Microsoft settle smart phone lawsuit. Infoworld http://www.infoworld.com/article/04/09/13/HNmssendo_1.html

Federal Highway Administration (1993) http://www.fhwa.dot.gov/ohim/1994/section7/in.pdf

Funk J (2001) *The Mobile Internet: How Japan Dialled Up and The West Disconnected*. ISI Publications Limited.

ITU Report (2002) Internet for mobile generation. http://www.itu.int/osg/spu/publications/sales/mobileinternet/

Luna (2002) Lukewarm take rates reported for i-mode in Europe. *Wireless Review*. http://wireless-review.com/ar/telecom_lukewarm_rates_reported/

Matsunaga M (2002) *The Birth of i-mode: An Analogue Account of the Mobile Internet*. Chuang Yi Publishing, Singapore.

Mobile Media Japan (2004) http://www.mobilemediajapan.com/

Shahin J, Heinonen A and Terzis G (2002) Mobile news study. Paper presented at *Mudia Conference*. Brussels, June.

Sigurdson J (2001) WAP: OFF – Origin, failure, future. Paper prepared for *JETS*. www.telecomvisions.com/articles/pdf/wap-off.pdf

Tee R (2002) Rijzende zon of donkere wolken? Mobiel Internet: leren van Japans succes i-mode en van fiasco WAP. http://www.computable.nl/artikels/archief2/d08ra2yy.htm

Van der Molen (2002) In de Week 17 – 2002. Computable, 26 April 2002, nr 17, page 23. VNU Publications. http://www.computable.nl/artikels/archief2/d17ra2qq.htm

Van Impe (2003) Vodafone reports 1 million live! Users, Mobile Commerce Net. http://www.mobile.seitti.com/story.php?story_id=2928

Volkskrant (2002) KPN moet groeien door nieuwe diensten (KPN should grow by introducing new services) Volkskrant, 21-12-2002.

Williams R, Edge D (1996) What is the Social Shaping of Technology? http://www.rcss.ed.ac.uk/technology/SSTRP.html

Wyatt S (1998) Technology's Arrow – Developing information networks for public administration in Britain and the United States. UP Maastricht.

Glossary

2.5G   The generation between 2G and 3G, with speeds enhanced by, for example, GPRS.

3G   Third generation mobile telephone systems that will combine voice and high-speed data services and offer a wide range of multimedia services when fully developed. W-CDMA as part of generic UMTS making this possible.

ARPU   Average Revenue Per Unit. One indicator of a wireless business operating performance. ARPU measures the average monthly revenue generated for each customer unit.

cHTML   Compact Hypertext Markup Language. The language in which NTT DoCoMo developed its i-mode services.

Content Provider   An enterprise whose products are information based, that is content, owned or managed for third parties. Content providers often include services to access and manage content.

EZweb   Mobile Internet services offered by KDDI, based on the WAP protocol, have been able to achieve the same penetration in its customer base, as has DoCoMo for its i-mode.

GPRS   General Package Radio Service. An extension for adding faster data transmission speed to GSM networks. GPRS is a package-based technology. Not to be confused with GPS.

GPS   Global Positioning System. A system, which uses a satellite to confirm a user's position on the earth surface.

GSM   Global System for Mobile Communication. The European Telecommunications Standardisation Institute (ETSI) and various EU research programs, such as RACE, played an important role in establishing this standard.

HTML Hypertext Markup Language, the language in which web pages are presently created.

i-mode NTT DoCoMo's Japanese Mobile Internet service.

Java A high-level object-oriented language, allowing applications to be written once, run anywhere, whatever the platform.

J-Phone One of the three large mobile operators in Japan in which Vodafone has taken a large equity stake.

KDDI One of the three mobile operators in Japan. KDDI is offering two mobile services, TU-KA and Au, with EZweb being the mobile Internet service that is offered to Au customers.

MML Mobile Markup Language used by J-Phone, one of the three mobile operators in Japan, in its mobile Internet service – J-Skywalker.

NTT DoCoMo The largest mobile operator in Japan, still member of the NTT Group that maintains a majority equity stake in NTT DoCoMo.

SMS Short Message Service. Method within the GSM-telephony for sending short messages from and to mobile phones.

UMTS Universal Mobile Communications System. The 3D mobile phone system.

W-LAN Wireless Local Area Network, based on the 802.11b IEEE protocol. Regarded as an infrastructure that has the potential to compete with 3G networks by offering local access users of mobile devices.

9

Context Perspectives for Scenarios and Research Development in Mobile Systems

James Stewart

9.1 Introduction

There is a huge amount of technical research and business investment in wireless and mobile technology that could be the basis of future systems available to the public. The current generation of mass-market technology primarily supports one-to-one voice calls and simple messaging. Newer technologies, such as Wireless Local Area Networks (WLAN) and third generation (3G) cellular systems most notably, offer the capability of many more services, from Mobile Internet through video messaging to wireless gaming. Mobile digital television and picture enhanced radio is set to become available in some form by 2006. Departing from the traditional mobile phone, the number of devices that can be used wirelessly is also the subject of considerable innovation; Personal Digital Assistants (PDAs), smart phones, tablet computers, wireless game terminals, plus many "fixed" devices. However, there are still many uncertainties about who the users might be, how and where these services may be used, what infrastructures and technologies will be built to provide them and who will make profits.

This chapter is based on work undertaken as part of a technical research project, FLOWS (Flexible Convergence of Wireless Standards and Services), aimed at creating a sophisticated technical integration of a range of wireless systems on a single terminal. The engineers needed models and scenarios for use of this technology that are based on research on existing use of mobile telephones, computers and other information and communication technologies (ICTs), and visions of future use. A method was developed to make existing social science research relevant to the engineers, and in the process develop a framework for imagining future users, patterns of use and appropriation and directing user research. It proposes some key insights into the development and use of scenarios in this process, highlighting weaknesses in existing scenario development.

"Integration" terminals form an essential part of the vision of convergent network systems that is very powerful within the telecommunications and information technology (IT) industries. Interoperability between different networks and radio systems – Radio or WLAN, Universal Mobile Telephony Standard (UMTS), Digital Video Broadcasting (DVB), fixed wireless and fourth generation (4D) radio access networks being the dominant – is seen as a way of exploiting the different characteristics of each technology: their bandwidth, mobility characteristics, quality of service, coverage, etc. All communication and "content" could potentially be carried over all radio access networks with an Internet Protocol (IP) core network using common protocols. Full network interconnection on the model of Global System for Mobiles (GSM) and the Internet should bring huge benefits from "network externalities". "Integration" terminals that can connect to several different radio networks are a fundamental part of a number of technical visions, such as "always best connected" when the most appropriate connection is chosen depending on availability, price and use, or simultaneous use, the FLOWS concept, where several radio networks are used at the same time.

However, it is not just a matter of technical possibilities. This convergence could be used in the realisation of two dominant industry visions or poles of attraction: the "Mobile Internet" vision and the "rich voice" wireless world vision; one being driven by the IT industry (predominantly US-centred) and the other, the mobile telephony industry (European and Japan), respectively (Lehr and McKnight, 2002; Stewart *et al.*, 2003). Competitive pressures and alternative visions are shaping future wireless, and present many technical and commercial challenges to convergence. For example, there is considerable resistance among operators to creating one terminal that will provide mobile telephony using licensed spectrum, and an Internet connection using licensed-exempt bands on the other. This represents a struggle between operators, vendors, content service providers or software platform firms to control, configure and brand the terminal. The challenge is all the more difficult given a climate of uncertainty over what people might actually want to pay for, corporate debt, increased competition, changes in spectrum regulation, new technologies and a general desire by corporate and mass-market customers to control, if not reduce, costs.

This chapter is primarily about developing tools to use social science research as inputs to specific scenarios that engineers will use to test and design a specific technical configuration. In the case given as an example, this configuration assumes a range of WLAN, UMTS and GSM services that will be available under particular physical, user and business circumstances. The scenarios must therefore imagine a world not too different to today's in which we are expected to have and to use a greater range of wireless connected devices and services, and highlight everyday situations that would test this technology in various ways.

9.2 Some Definitions

Communication in cross-disciplinary research is hindered by different terminologies amongst various specialist groups, particularly where the same term (especially service, application, scenario and user) has different meanings for different groups. There is no immediate prospect of aligning these usages, which are deeply embedded within the expert communities, such as telecommunications engineers, computer scientists and social scientists. In an attempt to aid clarity and communication in the FLOWS project, a number of more specific terms were developed and defined. The definitions are not intended to be exhaustive and are presented to offer some consistency in an area beset by very broad and often competing uses of the terminology. Of particular difficulty, as the UMTS Forum (2002) points out and fails fully to deal with, are the terms "service" and "application", which are explored in more depth and redefined.

Instead of the generic term "service", which can mean anything provided by one party to another according to a contract, the FLOWS project used the term Network Service, a telecommunications transmission facility optimised for particular Data Applications, such as speech telephony, asynchronous data transfer, etc., but not limited to those uses (e.g. analogue speech telephony services can also be used for data transmission). These can be provided over a range of bearer systems, such as GSM or 3G mobile network, a local wireless network or digital broadcasting, with some systems being more appropriate than others. The term Service Product is used to describe a combination of these that can be sold to end users or resellers. For example, mobile Intranet access for laptops is currently being sold in a product that uses WLAN, GSM and 3G to provide data-only access. Some phones can switch between local wireless and a mobile system, and still provide voice calling and text messaging.

Further, three terms were coined relating to the idea of application: Application Domain, Application Package and Data Application. These definitions make it easier to distinguish what is being discussed and link user and market-based definitions to technical definitions (Figure 9.1).

9.2.1 Application Domains

The term Application Domain links a domain of activities, such as education and learning, shopping/retail, leisure, community participation (the "what") with the enabling technologies or processes that is the "how" (for

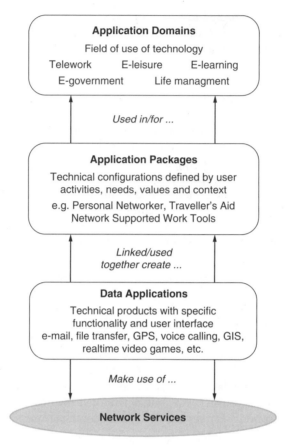

Figure 9.1 Three definitions of application.

instance "e-", "tele-"), creating applications such as "e-learning". Applications understood in this sense then follow those presented in the EC IST Beyond 3G Cluster Report (Beyond 3G Project Cluster, 2002). In creating analyses and scenarios of business and use, the general term "application" is used to refer to what people are using the technology for. The classes of application referred to in this case depend on the type of analysis being made; for example, they can be grouped according to functional activities or market sector (e-learning, teleworking, e-leisure) or according to cognitive and social use (belonging, control, intimacy, etc.).

9.2.2 Data Applications

Data Applications are specific data or information handling applications based on use of computing and communication services, the use of the term application commonly used to refer to computer programs used by end users. Data Applications have a narrow function or specification, both from technical and enduser points of view, but are used to support a range of user activities: e-mail can be used for e-commerce or tele-education. A Data Application is a combination of software and hardware that directly interacts with a user, and is thus defined from a user point of view and a data processing perspective. Their description is essentially functional. Familiar Data Applications include e-mail, word processing, FTP, web browsing, short messaging, voice telephony, P2P messaging, GPS, video on-demand, etc. There are many more specialist Data Applications designed for particular industries, including mobile surveillance, urban guidance, tourist guides, parcel tracking, etc. (Velez and Correia, 2001). These Data Applications generate particular demands on service infrastructure, utilising one or more types of network services.

Network Data Applications generally require technology installed in the users' device and that of their correspondent in the case of telephony, or application service provider (e.g. e-mail accounts, on-line banking, file server). In telephony and Internet provision, the network service providers generally provide only basic additional applications, such as voice mail or e-mail (and Short Messaging Service, SMS) servers. All other applications are accessed on third-party providers' technology, using the user's device.

The Data Applications that we are familiar with are being reshaped in meaning over time and with technical changes. For example, e-mail can be implemented in different ways, and to the end user, appear to merge with multimedia messaging. A web browser can be used to read e-mail, and the sending and downloading of larger and larger attachments means that it is as much about large file transfer as short message reading. Many specialist applications are increasingly based on the combinations of a limited set of "standard" Data Applications (Williams *et al.*, 2005) using common standards. The way Data Application is defined is very much related to current practice and available technology. As these change our perception of basic data functions changes too. What is at one time a peripheral Data Application, seen as a subsection of another application, may become in time a dominant or standard application and vice verse. It is well worth spending some time considering what the basic underlying data transfer operations associated with existing, and perhaps future, applications are.

To be used in the development of scenarios, Data Applications are defined not only in general terms of what they provide for the user, but more importantly in strictly limited terms that enable simulations of use in engineering models. The quantitative description of each Data Application is often a theoretical model, rather than an empirically derived, statistical model. This is particularly so since we are dealing with future uses of applications that are largely unknown. What FLOWS elected to do was to select a small number of well-known applications that represent a sufficient range of demands on network services to be a useful model.

9.2.3 Application Packages

As well as the meanings of Application Domain and Data Application, there is another use of the word "application" that refers to how people use a combination of a number of Data Applications together with various network and content services. This combination or package is optimised, or at least suitable, to support activities within an Application Domain (i.e. reflect specific user requirements). These groups of Data Applications can be called Application Packages. In other words, Application Packages use a number of Data Applications to undertake particular tasks, or to support particular activities, relationships, etc., that are identified in the user scenarios and research on the environment of use. An Application Package has a description based on what someone uses a device and Service Product for and the symbolic meaning they give it. It has a much richer description than the functional description given to a Data Application.

The concept of the Application Package is important when the actual Data Applications that may be available in the future are in some degree uncertain. Application Packages are identified from a user perspective (Cook and Aftelak, 2001; Hickey and Pulli, 2001; Cook, 2002). They describe what a user wants from a technology and the role it plays in their everyday life. The concept is critical in the development of intelligent systems on which the user can set preferences and establish a profile leading to greater personalisation. The names given to them reflect what they do from the user's perspective, not from the perspective of the types of data exchanged: for example, a City Survival Kit describes how many people use their mobile phone today, a use that can certainly be enhanced considerably with new technology.

Any device is likely to contain a number of different Data Applications that could be configured together in different ways to produce different packages; for example, a single device could be used for teleworking, maintaining family intimacy and media consumption. The existence of different packages or configurations on that device reflects distinct and separated roles in a person's life (Nippert-Eng, 1995; Gournay and Mercier, 1998). However, there are an increasing number of packages that enable people to integrate different parts of their life, as well as keeping them apart.

Applications Packages can be strictly defined by application providers and controllers, or shaped by users responding to pricing, quality of service, etc. with particular patterns of use. Technology producers attempt to link together a number Data Applications into a single package (a PC "Office" suite), but also recognise that standards-based interfaces (e.g. web browser, operating system desktop) enable users to configure their own package of Data Applications.

Application Packages are generally supported by a "matching" Service Product that provides all the Network Services and Management that the packages require. However, a user may not find a suitable product from any one supplier, so will buy or obtain from several sources, for example, WLAN roaming and GSM from different operators. The value added by the application and device producers is in creating a product that can manage, and switch between the Service Products of the different network suppliers (e.g. UMTS from one supplier and WLAN from another).

9.3 The Socio-technical Approach

The basis of much of technology studies is the idea of the "socio-technical". Devices, Services and Applications are not just technical, but socio-technical constructs embodied in infrastructure investment, devices, service agreements, business models and use conditions. The availability and development of technologies are not based solely on what is possible, but what people and organisations decide to make possible (Williams and Edge, 1996; Williams *et al.*, 2005).

An example from mobile telephony is the GSM phone service. The user does not simply buy a device to use voice telephony, but needs access to a range of technologies, services, commercial contracts, social norms, legal rights and restrictions, etc. The service "package" is shown in Table 9.1.

The particular combination of factors shapes how the application is used by the user, for example constant calling, emergency calling, SMS dominating

Table 9.1 The socio-technical mobile phone package

Network Service Product	Extended device and Application Package	Use factors
Wireless speech service + connection service + location service Network coverage (physical access to the radio network)	Phone device Charger + electricity	Social rules on appropriate use of mobile phones: locations of use, topics of conversation, sharing, messaging, etc.
Service agreement (contact between user and service provider) Payment method/agreement – prepay, pay monthly, etc.	Net services – e.g. call diversion, voicemail, mailbox, SMS forwarding, Wireless Application Protocol (WAP) gateway (Network Data Applications)	Bill-payers rules on use: e.g. parents, corporate Personal usage patterns and meanings: personal rules, reliance, habits, etc.
Time/price/minutes/subsidy and contract	Subsidy – related to integration with Service Product	
Roaming agreements between networks (national and international)	Insurance for loss or damage	
Wireless device operation license (granted to operator) Customer Support Services	Customer Services	

over voice calls, always having the phone on, etc., whether it is suited to the Application Domains they are using it within and the Application Packages that they use. Another example of a social-technical definition of a service/application is Internet access. Although we have a common understanding of what this is, it is actually a range of possibilities: a set of possible network services based on a service agreement: pricing, allowed times of use, maximum bandwidth availability, quality of service or particular sub-services, allowed access locations and devices, various network services (mailboxes, caches, content filtering or digital rights or censored content, firewalls, etc.). It is therefore important to define precisely a range of factors, even when using an apparently neutral term.

Any new application, service or device has to be analysed within a socio-technical framework. This means user scenarios are socio-technical, describing not only social and personal activities, but also interaction with the physical world and the world of machines.

A socio-technical approach also treats technical change as a socio-technical process, not defined purely from the independent emergence of new technologies and techniques. This research is based on a social shaping perspective, that suggests technology is developed and deployed as the result of the complex interactions and actions of social actors including firms, governments and various user groups through market and other relationships. Even end users, with little direct influence over product and service development, adopt and domesticate new technologies in ways that can eventually have profound impact on the future direction of technological change. One may suggest that real or imaginary knowledge about users and customers is becoming one of the most important actors in the social shaping process today (Williams *et al.*, 2005). For this reason, understanding people's everyday lives, specific life events, and the way that people actively integrate technologies into their lives in practical and symbolic ways is very valuable to those making decisions concerning investments in technology R&D, deployment and marketing. It is also important to be critical of the sources of information about end users, and recognise the need to keep new knowledge about users flowing into any technology or service development throughout a project.

There are many ways to understanding users: studies of wireless technologies and of other information and communication technologies have yielded many different ways to view users, and the way they use technologies. This includes why they use them, how they use them, the motivations and problems they have and concepts, such as role, behaviour and community, etc. Recent books and collections (such as Brown *et al.*, 2001; Katz and Aakhus, 2002; Ling, 2004) on the use of mobiles serve as sources. One of the problems with looking at the future is that it has not happened yet, and mobile phones have only been available for such a short time. However, there is plenty of research on how people live their lives and work together, which changes much more slowly than the technology, so provides good grounds for speculation.

There are many ways to approach understanding how and why people use technology. Traditionally the concept of application domain or use has been based on some function that an industry sector provides (e.g. education, work, entertainment, shopping selling, leisure, communication, etc.). Alternatively, in studies of how people use ICTs it is common to base the analysis of types of information activity: information seeking, browsing, access, communication, transactions, media consumption and play. However, studies of users will often focus on other type of uses, understanding the concept of application in very different ways:

- knowledge, communication, service, play, verification;
- immediacy, intimacy, flexibility, freedom (closeness of human relations);
- belonging, playing, coping, survival, balancing, delivery, control, freedom (related to balancing activities and relationships).
- time saving, time wasting, time filling (related to time use).

The development of quality research and useful scenarios relies on bringing together a range of disparate perspectives: in this case the analysis of the application and use of technologies is greatly strengthened by using multiple frameworks.

9.4 The Context Perspective Framework

There are many ways of looking at the users and usages of technologies, and of mobile or wireless technologies in particular. Categories can be based on what individuals or communities of users use technologies for, on particular Application Packages, or on the places that wireless technologies are used. Users can be individuals, but also organisations that provide individuals with wireless technology, and the suppliers of technology infrastructure. However, the role of this chapter is to go beyond a person-centred approach and try to describe the context of use of telecommunications services. This entails describing the locations of use, and the devices and applications software used to access them. This approach draws on the sociological, ethnographical, cultural and geographical study of ICTs in general and mobiles in particular (Lie and Sorensen, 1997; Gournay and Mercier, 1998; Ling, 1999; Haddon, 2000; Grant and Kiesler, 2001; Haddon, 2001; Haddon *et al.*, 2001; Ling and Yttri, 2002; Katz and Aakhus, 2002; Ling, 2004), business/design literature on mobile futures (UMTS Forum, 2002, UMTS Forum, 2000b, UMTS Forum, 2000a, Cook, 2002; Cook and Aftelak, 2001, Dietrich and Eichner, 2001; Woodward *et al.*, 2001; Lehr and McKnight, 2002) and technical descriptions, for example in 3GPP (Third Generation Project Partnership) specifications (3GPP, 2003), or in work done in the FLOWS project redefining network services and application descriptions.

A Context Perspective Framework is a way to structure the bringing together of material for the creation of user scenarios. This approach reflects a number of different analytic perspectives on ICT use and the needs of this project. Within this framework four perspectives are used: location, people, device and application package. For each perspective a number of characteristics are developed, relevant to the characterisation of services (i.e. in relation to geographic space, number of users, Data Applications, demand for services, quality of service expected, etc.). Much of the quantitative data for this are available, although there are still large gaps in knowledge about what traffic patterns will actually be generated by users. There is also qualitative data, but still there are limits on how this can be translated into scales that can be used in scenarios useful to engineers. Within each perspective a number of examples or context scenarios identifying relevant characteristics are developed. This provides a framework not only to understand service requirements, but also to develop further in-depth user analysis.

The context perspectives are not mutually exclusive: users move through particular locations, using particular Application Packages, these Application Packages are used by a range of different users and locations are inhabited by a range of different users. This overlap strengthens the different scenarios by cross-referring and also helps to generate a large range of possible user scenarios more easily.

In order to identify the characteristics of the various network services, a Data Application (e.g. video streaming) is chosen and the information about where, how and who can be found easily from the scenario outline tables showing the range and constraints of possible uses of the application and the services needed to support it (Figure 9.2).

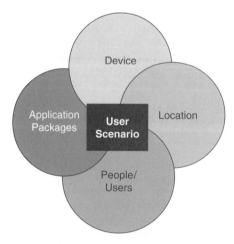

Figure 9.2 Inter-linking context perspectives.

9.4.1 Location Perspective

A key dimension to the use of mobile and wireless services is the location in which they are used. Mobile technology opens up many new spaces to the use of ICTs, and this perspective highlights particular spaces that people move through and to in the course of activities, especially those that are by definition related to travel and temporary occupation. The key characteristics of locations are: physical characteristics of the place, number and type of users occupying or passing through the space, time and space distribution of users, etc. the applications and services that they use, mobility within the location and limitations and the infrastructure deployment including cell size. For the social and commercial analysis there are characteristics of control of the space, competition and provision of wireless services, inter-regional and international roaming. Physical descriptions of locations can be obtained from measurements made in studies of radio propagation. Human use of locations is obtained from studies by urban planners and architects, specific studies, and details of current wireless use from mobile operating companies.

Examples of locations include airport, railway station, tourist city centre, industrial estates, suburban home, open plan office, school, shopping mall, high street, sports centre or stadium, exhibition centre. It can be extended to rather more restricted spaces such as a car or train too.

Locations can be quantified in terms of topography, geography and type of wireless cell availability, density and number of users, and typical use of wireless infrastructure (Stewart *et al.*, 2002).

Take, for example, the shopping mall. The shopping mall can be characterised as a space located in city centres and suburbs which attract large numbers of visitors, with varying levels of occupation, depending on time of day, day within the week and time of year (e.g. seasonal shopping such as Christmas). Shopping malls are perceived as centres of leisure and entertainment and a range of facilities, such as restaurants, cinemas and hairdressers are available in addition to the range of stores. Shopping malls have also become popular teenage hangouts. However, some people find them very stressful and try to spend the minimum time needed for shopping. Highly connected mobile devices for consumers are a shopkeeper's nightmare. They enable customers to compare prices, availability and design right in their store. However, wireless technologies could offer the opportunity for shopping centre and shop managers to add an extra dimension to the shopping and selling experience. (Table 9.2)

9.4.2 Users/People Perspective

The viewpoint of users, or more correctly people, focuses on the activities, community and resources of individuals or specific groups such as a family,

Table 9.2 Shopping Mall (Stewart, 2002)

Physical characteristics	Type of users	Data applications	Mobility	Device	Infrastructure
Indoor open space: with narrow corridors and enclosed shop spaces. Glass, concrete, steel construction	Families	Products information search (including price, location product within mall and ordering facilities)	Walking	Basic devices (phone, PDA)	High provision of infrastructure
	Teenagers		Browsing	PDA+ for stock takers, delivery	Public pico and microcell GSM and UMTS
Most areas publicly accessible, except for staff only areas (e.g. security control room, staff room, stock rooms, etc.)	Tourists		Sitting	Pen size (Ultra portable)	Public WLAN and private WLAN for shops
	Other shoppers (young professionals, students)	Secure payments systems	Vehicular (car park)	Laptop	Competing and locally provided cellular access
Outdoor/Indoor space: adjacent areas (e.g. parking, small parks)	Staff (commercial premises, maintenance, security)	Bank account access			Short range radio systems
	Business people	E-mail			
	Construction personnel	Multimedia messaging			
		News			
		Voice			
		Web browsing			
		Music			
		Games			

across their entire life space. It links the high level ideas of life themes and life projects to the everyday management of the life space, performance of activities, consumption and integration into community and organisations. Of course every individual has their own pattern of usage, but it is possible to suggest a number of example types of user around which to build a scenario of needs, behaviour, resources, etc. Possible user types include: teenage school child, student on campus, high level travelling business executive, stressed commuter in large city, service engineer, or regional sales manager. Users are chosen from a range of people identified in the literature as having particular uses of information and communications, and as being especially "mobile". The scenarios can be based on some sort of functional description of the person, for example, their job or life stage, or on some other characteristic. They are not intended to be based on existing proposed classification of technology users, but could be based on various other lifestyle or marketing categories. The greater the number and greater the variety of examples to draw on in later developing user scenarios the better.

In this case a heterogeneous range of people with different resources, activities and needs was chosen. In doing this, an attempt was made to balance selection criteria based on traditional demographic factors, lifestyle factors and groups identified in the literature on ICTs and mobile telephony. The relevant characteristics of the user suggested are location, mobility, Application Package, affordability/resources (including cost of services, devices and support), flexibility (level of service expectation, e.g. the need for constant availability), including the user profile and devices used. The social and economic themes and characteristics include issues of freedom, independence, control, surveillance, coping, identity and community. The ways to measure and quantify some of the characteristics certainly need future work, but they provide a starting point.

The commuter is a common example in the mobile literature. The suburban commuter uses wireless devices extensively while travelling to and from work during the week, as well as during excursions at the weekend. The offerings of the technology are an essential and integral part of these daily trips, to keep in touch with work and personal network. It also provides distractions from the boredom, stress and routine of travelling, by making use of this time. Themes include coping with the stress of life, balancing home and work, maintaining personal control and managing family life (Gournay and Mercier, 1998; Kopomaa, 1999; Haddon et al., 2001; Crisp et al., 2002) (Table 9.3).

9.4.3 Application Package Perspective

The Application Package relates to the particular activities that a user is engaged in and the set of Data Applications and services that are relevant to that activity. The concept of Application Package links the activities of users to the use of the technology. Application Packages refer to particular activities,

Table 9.3 Commuter

Location	Mobility importance (1–5)	Data application	Total affordability	Flexibility in access and reliability	Dependency	Device	Application packages
Train, bus (i.e. public transport)	Fixed (3)	E-mail	Medium	Medium	Emotional: high	PDA	Personal Networker (heavy)
Car	Pedestrian (4)	File transfer				Mobile	
Office	Vehicular (4)	Voice call			Practical: medium–high	Fixed PC at work and PC home possibly other terminal type)	City Survival Kit (heavy)
Supermarket	Highly vehicular (4)	Time-killing games					Media Consumption
Shopping mall		Bulletin boards					
Pub		Web browsing: product information					Portal (Heavy–moderate)
Home		News					NSW: mobile office, Intranet (moderate)
Lives within the greater city area and moves around public transport routes to centre and between suburban locations		Travel information access					
		Video clips					
		Music					

such as education, work, leisure, life management, but in a rather more focused way than Application Domain. The Application Package scenarios identified include several types of Network Supported Work: the Mobile Office, Corporate Intranet Connection and Remote Monitoring/Control system, the "Traveller's Aid", the "City Survival Kit", the "Media Consumption Portal" and the "Personal Networker".

Characteristics of applications identified are: information and communication tasks, Data Applications and usage (time and location). The personal social and economic characteristics include: immediacy, cost, competition in provision, control over provision and use, transferability between devices and reliability.

Here four different application packages are illustrated; three that can be classed as network supported work tools and one entitled the "City Survival Kit":

Teleworking has traditionally been seen as working from home, but is increasingly recognised as any sort of peripatetic work; people working out of a car, or truck, plane or tractor, or indeed anywhere (airport, building site, oil rig, other people's homes, etc.). Application Packages for telework have been termed Network Supported Work Tools (after the term Computer Supported Cooperative Work, CSCW) and include applications for a range of different types of work. Work on CSCW in mobile environments has been developing ideas in this area of several years (e.g. Dix and Beale, 1996a, b). These tools give flexibility and enable workers to be in contact onsite, or available wherever they are, connected to their organisation's management and communications systems. This enables them to receive instructions, submit information on work progress, query colleagues and databases, or monitor remote systems. This can mean considerable efficiency and effectiveness gains, and reduction in costs. It may mean working at home but being in contact with colleagues at work and with customers through video/voice/data sessions (virtual office). It also means collaboration between geographically separated persons, possibly a group of them. Here too, the ability of telecommunications to deliver video and sound as well as real-time data may allow users to avoid costly and time-consuming travel. Application developers have caught on to this opportunity. A variety of screen sharing tools provide users with the means to work together in real time on the same electronic documents while being in visual and audio contact. These tools not only enable people to work, but also undertake training and juggle their work and home lives. (Gournay and Mercier, 1998).

Here are three brief examples of different network supported work tools:

Intranet Connection: For the Mobile enterprise, the wireless link is seen as an extension of the corporate Intranet, with corresponding expectations on availability and bandwidth. The application enables access to a network, to take out and feed in communications and information. This package enables people to maintain a presence in a virtual office and organisations to deploy "Mobile Enterprise Systems". It offers a rich variety of office type Data Applications from voice telephony to document sharing.

Mobile Office: Rather than being a remote mobile part of an Intranet, the mobile office acts as an information and communications' hub for its user, enabling them, for example, to run a business from wherever they are (Laurier, 2001). The independent worker uses network-based applications (voice, mail and web, and other application servers) as well as portable devices.

Remote monitoring/control system: Here the mobile device is often connected to another mobile device (e.g. diagnostic tool, surveillance device or other remote machine such as a health monitor, car engine monitor) to enable this system to communicate to a remote human or machine (e.g. remote surgery, engine diagnostics, video surveillance). The remote device can be controlled by a co-located person, remote person or be independent part of a machine-to-machine network. Ubiquity of coverage may be more important than bandwidth availability. A system often used by a "Clipboard Technician".

Take, for example, the 'City Survival Kit'. A common use of mobile technology is to cope with the stresses of living and working in a busy city where every activity involves travelling and dealing with the uncertainties of modern life. This package is chosen by the user who is often on the move, trying to make rendezvous, dealing with public transport delays, the demands of work, family and friends, paying bills and ordering shopping while travelling, finding out what's on, what is available and where things are. Location-based applications could be important (Kopomaa, 1999; Sherry and Salvador, 2002; Townsend, 2002,). Clearly this is the sort of package that could be popular with commuters and city dwellers.

9.4.4 Device Perspective

There are a number of core mobile devices; the phone, the car and the PC being the dominant today. But there is a great deal of innovation going into developing devices with a range of forms, encapsulating different technologies and for different uses. There are a number of possible devices that services can connect to, that applications run on and which offer different levels of portability. The characteristics, such as form factor, (size, shape), power source and Application Packages of a device shape the possible uses, users, usages and connectivity. Of key importance are not only the stand-alone capabilities of the device, but the degree to which it can interconnect with other wireless and wired devices. Many Application Packages will run across a range of devices and most users will have access to several devices simultaneously or in different parts of the life space (e.g. the phone, laptop or PDA) can always be updated from a computer and fixed wire link at home or work on a regular basis, ideal for many asynchronous solutions. Users may increasingly expect to be able to switch a call or data session from one device to another, or conduct different parts of it over several, not only in the familiar user-configured way (e.g. using phone and playing an online game with friends), but through an integrated communications system. In its most integrated form this has been termed the virtual terminal (Thank *et al.*, 2001). A number of types of devices (or device scenarios) have been identified in (Table 9.4).

Characteristics of devices include: portability/wearability (Fortunati, 2002), human interface, modes of communication (text, voice, image, etc.), cost of ownership and use, power requirements, processing power, range of Data Applications, locations it can be used, social acceptability, etc. Location and device will have a shaping effect on what can be used, or is inappropriate to use (e.g. sending text message/e-mail in library, but not talking on phone). Locations imply some restrictions on type of use as well as use of particular devices in general, for example, a laptop can be used on a train, but not the metro at peak hours (Table 9.5).

Take for example, the PDA+. The PDA+ or Web Tablet format is larger than a standard PDA and also much heavier on features. In 2004, Web Tablets with large screens were rare, but the PDA+ is a widespread format in commercial use (warehouses, logistics, retailing, utility companies, highway maintenance, etc.). These products tend to be used by corporate/commercial users, "clipboard technicians", who find the portability of heavier-featured devices particularly appropriate. Cheaper and more fashion-oriented designs of the device may appeal particularly to the youth market and chatterers, with features and uses centred on MP3, media consumption (games, video clips) and personal network applications, where the device offers a larger screen and longer battery life than smaller devices. Often based on a PDA size pocket computer device it can be enhanced with features, such as a large screen, printer, a camera, larger battery, a small text or music keyboard, hard disk drive, or specialist modules (Table 9.6).

9.5 Scenarios

The aim of scenario development is to create models for testing, design, simulation and market studies, and a common framework for all the team members to work within. Scenarios must be built with a clear set of inputs and variables that can be changed in a way that different versions of a scenario can be tested or investigated. Assumptions, uncertainties and gaps in knowledge should be stated and sources of inputs provided.

The scenarios developed here enable the linking of research on people to visions of possible future wireless devices, to quantitative descriptions of a number of Data Applications and Network Services. This is done through User Scenarios developed from a combination of the context scenarios presented earlier. Context scenarios of individuals, applications, location and

Table 9.4 City Survival Kit

Uses	Data application	Key Values	Usage	Device
Keeping up with social network, and work: messaging, bulletin boards, phone	Web browsing	Coping	On public transport or in the car, but also in places en-route.	Basic devices (Phone, PDA, Laptop)
Transport info: train times, traffic alerts	E-mail messaging	Freedom	Used in work and personal time	Extended basic mobile phone.
Terrorist/accident alerts	Voice call	Control		Maxi-mobile (pocket size but heavier features)
Shopping	Instant multimedia messaging	Awareness		Pen size (Ultraportable)
Paying bills	Games	Time saving		Emphasis on portability and design with some extra functionality
Local timely information: what's on	Video on demand			
Games – time filling and wasting				
Music (listening and purchasing) – time filling				
News				
Television, movies and radio (live, on-demand or prerecorded)				

Table 9.5 Device scenarios

Size/name	Description
Basic formats	
Mobile phone	Portability, low-end memory and processing power applications and application packages
PDA size	Portability, multimedia data communication. Many possible applications and packages
Laptop	Full computing and interface power, many applications
Extended Basic Schema	
Mini-mobile/MP3 player	Emphasis on portability (<100 g) for basic communications and info, or music playing. Limited by screen size and battery size
Maxi-mobile/Smart Phone	Jacket pocket or handbag size, but heavier on features (big screen, PDA, camera, etc.)
PDA/GameBoy size	Palm size considerable computing power, carried in bag, some pockets. Clip on keyboards, etc.
Web Tablet or PDA+	A5–A4 size. Tablet, PDA with extra devices attached (e.g. GPS, printer), too big to go in a pocket, but can be carried by hand or in a bag
Mini-laptop	Scaled down laptop, with keyboard – goes in small bag, many features of PC laptop: bag size, full power, keyboard
Laptop computer	Full-featured computer with large screen, multiple ports, disks, etc. Luggable
Other Formats	
Vehicle based	Provides more powerful communications and IT facilities in a mobile location. Provides a hub for portable devices
Communication hub	Communications device that is carried separately to other devices, possible to get better signal or to avoid radiation. This could be small or built into briefcase, etc.
Clothing based	Integration into clothing links telecommunications intimately with the wearing of that clothing (especially work clothing)
Badge	Wearable delivering context awareness and presence
Pen or watch size	Ultra portable

device are linked together and developed in a rich and more specific narrative and description. This could of course be largely fictional, or based on a particular set of research findings. For example, a scenario developed around a building site would link information about all those who come into the site and communication needs of key users as they move between the site and other locations.

However, user scenarios alone are not sufficient for use in system or service development. There must also be input on definition of network services that can be provided and on the infrastructure that will provide this. One way to do this is to use Service Scenarios and Access Technology Scenarios

Table 9.6 Web Tablet/PDA +

Characteristics	Mobility	Size/weight	User	Location	Application package or Specific Data Applications	Performance	Support network	Extended features
Portable	Fixed (5)	200 × 100 × 50	Mobile workers	Indoor/outdoor	Network Supported Worktools	High level and processing power for business/heavier featured devices	WLAN	Bluetooth headset
Built-in antenna or fold out slate	Pedestrian (5)	900–1500 g	(Service engineer, Peripatetic manager, etc.)	Urban/rural	Personal Networker	Standby time up 72 hours	GSM/GPRS	Notebook
Context aware	Vehicular (5)			For example, Industrial estate, airport	City Survival Kit		Bluetooth (PAN)	Desktop computer
Touch screen, speech, handwriting input	Highly vehicular (4)		Business travellers	Tourist city centre, suburban	Traveller's Aid		UMTS	Car (docking station)
Detachable keyboard			Teenagers Chatterers	Estate, financial district, shopping mall, public/private transport	Media consumption portal		DxB	Other peripherals (printers, etc, can be built into this format
Screen size (up to approximately 11 inch)			Tourist					

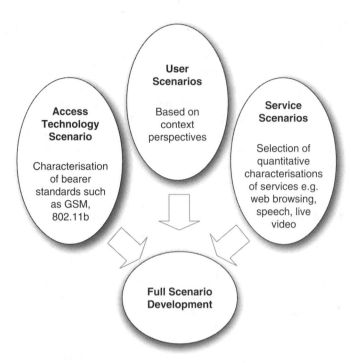

Figure 9.3 Combination of different types of scenario inputs.

that can be brought together with user scenarios developed using the context perspective approach (Aguiar *et al.*, 2002) (Figure 9.3).

9.5.1 Network Service Scenarios and Access Technology Scenarios

The Network Service Scenario describes the demands on the device technology and networks by users making use of particular Data Applications while going about their everyday activities. These descriptions include both empirical measurements and derived models. They are principally technical scenarios that define and describe a set of a network services. Service Scenarios involve the choice and characterisation of a range of Data Applications and Network Services, providing a detailed description of the parameters that define them. The Service Scenario describes the service capabilities that the network and the devices will be required to deliver. These parameters are grouped into three categories: bearer characteristics, traffic characteristics and performance requirements and include a number of services, quality of service, bandwidth/data rates, connection orientation, symmetry, security, switching between bearer services, etc. It will indicate what combination of access technologies could provide these services, what functionality is needed to support simultaneous use of these technologies

181

and the limits on use that these access technologies impose for defined levels of infrastructure availability. The scenario will include services, such as voice telephony, file transfer, services for real-time gaming, etc.

Access Technology Scenarios describe the bearer standards that could be used to provide network services to users. In the current wireless world these include various WLAN systems, Bluetooth, GSM and UMTS, but a future scenario may also include various 4G or alternative systems. A scenario describes the combination of standards, available, which helps determine what network services might be able to be provided, the number of users that might be supported and the types of mobility possible.

9.5.2 Creating Scenarios

Full scenarios are created by bringing together User Scenarios, the Service Scenario and the Access Technology Scenario. However, a single unified scenario is not sufficient for the needs of a diverse development team: engineers working on networks and those working on radio access have very different needs. Therefore a number of different, related, scenarios have to be built, which is why the development of a framework is so important. In this project two different sorts of scenarios were developed; those related to the network demands of particular users (i.e. User-Centred Scenarios) and those related to the aggregate demands of users in a particular location and the propagation characteristics of that location (i.e. Service Provision Scenarios).

For example, a User Scenario with a builder calling from a site discussing and modifying plans with the architect using a virtual workspace package with access to WLAN and UMTS networks is linked to the Service Scenario that includes a characterisation of a voice and data, video call with particular network infrastructure and technologies described in the Access Technology Scenario. A Service Provision Scenario describes the use of wireless services by all those in the building site, the physical characteristics of the space, the access infrastructure and the social relations and control.

9.5.3 User-Centred Scenarios

These are scenarios based on individuals moving though different places over time, using a range of devices and applications. They are used by engineers and planners working on devices and higher levels of communication, for example, hand over between different bearer systems, device power requirements or requirements of specific end-user markets.

Just as the User Scenarios describe more than one particular activity, a complete User-Centred Scenario is a description of all the different network services across a range of locations that a person or group makes use of in the user scenario. This could include their home, their car (driving around a town or motorways), their office or school, visits to friends, shops, etc. It describes what they do in each place, who they communicate with, the technical facilities available, the environment, the social organisation of each place and the position of the person within that organisation, and many more factors as are deemed relevant for the particular use of the scenario. These User-Centred Scenarios are the basis of Service Product Scenarios, developed later.

9.5.4 Service Provision Scenarios

Service Provision Scenarios are needed for developers and researchers focusing on locations, developing channel models or calculating the provision of radio access points that will be needed in a particular place. They require Service Provision Scenarios focused on place, rather than individual users or applications, aggregating all the users in that location. This includes information from the user scenario, such as number of users (given by per cell or per km^2), spatial distribution of the users, position and mobility of users, services used by users, traffic models for those users and details from the location such as available access technologies, geographical and topographical descriptions to create a propagation scenario. This type of scenario is a richer version of a Deployment scenario (Velez and Correia, 2000). Those studying social aspects of sites will want information on control of the space, social rules, activities and relationships in those sites.

For example, these service provision scenarios could be built on user scenarios such as:

- Fifty people in an small airport lounge uploading files from their laptop computers and PDAs within 5-minute period via WLAN network, but with UMTS via picocells also available; making voice calls from phones and browsing for travel information on PDAs.
- One hundred people making phone calls, 20 people playing on-line games, 150 sending and receiving small multimedia messages within a school playground with UMTS microcell and WLAN picocell.

For the FLOWS research on propagation, five locations were identified for testing (office, town square, enclosed corridor, suburban and rural) and the service provision scenarios were based around actual measurements taken in examples of these locations.

In general these scenarios are still too broad for most testing and development, so a series of quantified use cases or scenario instances have to be developed. These describe specific activities, and it is here that the scenarios have to be most explicit. Use cases have to be developed for the specific needs of the scenario user, and in consultation with them. Changes they make need to be fed back into the main scenario. In the case of our project, FLOWS, these instances needed to highlight the demands of simultaneous use of multiple bearer services.

Example Scenario Instances include:

- Making a multimedia call on a PC while travelling in a train at 120 km/h in an area of macrocell coverage using UMTS.
- Switching from a narrowband service to a broadband service during an interactive web browsing session on a Web Tablet while stationary in an area with picocell coverage (e.g. GSM to WLAN or UMTS).
- A streaming video service switching between broadband bearer services with a change of mobility from stationary to 30 km/h travel in an urban area (e.g. WLAN to UMTS).
- Switching an interactive document editing session from a stationary laptop PC with WLAN connection to an outdoor pedestrian using a PDA with UMTS service.
- Playing a real-time videogame on a GameBoy with average bit rate of 50 kb/s in a bus averaging 30 km/h over UMTS.
- Making a GSM call on maxi-phone while downloading a series of 30 kB images via GRPS.
- Switching broadband service due to congestion on one, in an airport (e.g. WLAN to UMTS).
- An interactive e-mail session via GRPS on a train that delays downloading large documents until WLAN contact in a station.

9.5.5 Understanding Scenario Connections

Figure 9.4 shows how a set of scenarios are constructed from Context Scenarios and the Service and Access Technology Scenarios. First, a User Scenario is created by combining a number of different context scenarios describing the four key dimensions of end user use of wireless technology: location, Application Package, user/person and device. In this case the user scenarios are centred around a particular user, who moves though several places, using different devices and Application Packages. This User Scenario in used in conjunction with the Service Scenario to describe a User-Centred Scenario: all the services and conditions that the user is described as using. Particular instances of use of convergence technology are highlighted as Convergence Scenarios. Another Scenario, this time describing the Service Provision requirements for a particular location is also generated, based on the User scenario, but this time centred on a particular location within that scenario. It aggregates the users, devices and applications used within that location.

There are yet more scenarios that this technique could be applied to: device-centred and application-package-centred scenarios would be useful in defining and testing the range of conditions under which a particular device or Application Package could be used.

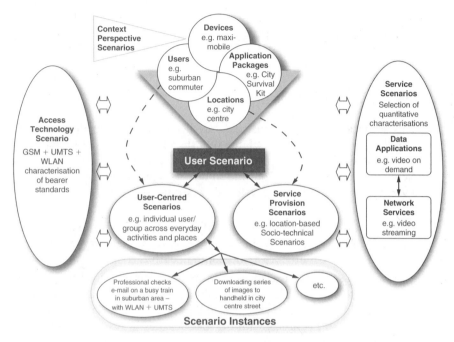

Figure 9.4   Scenario development.

9.6 Example Scenario Development

A "story" scenario is common within the industry and design methods. Sometimes they are based on purely tacit knowledge of engineers set up to test particular technological configuration, or as a way of illustrating the results of foresight scenario building exercises or built from specific user research. These narratives bring the research data to life in an engaging manner, but at the same time they can be dangerously alluring and justify the logic of a technology. They can also be taken out of context, as somehow being an important part of a scenario, rather than part of a process and a bridge to the original research. Three stories were created to illustrate different types of uses of the technology under assessment, and a range of people and places. In many ways they are quite conventional scenarios, the male business traveller, a family on a weekend trip to the city and a self-employed peripatetic service engineer. The careful reader will notice that broad environmental trends are hinted at, but these examples made no attempt to engage with these seriously as has been done in many other top-down scenarios developments.

For each scenario, the scene is first set by bringing together the information already gathered in the context perspectives and adding some more specific ideas. Scenario Instances are developed from this. The examples given were developed for this particular project's needs, to test the idea of simultaneous use of standards, concurrent use or sequential switching between different radio systems. Here one of the scenarios is given.

9.6.1 Business Traveller on the Train

This scenario describes a business traveller on a medium distance train travelling from the city, through the suburbs and rural areas, stopping regularly at stations. The main user is conducting business activities with a Network Supported Work Tool and some personal communications. This is a very typical example, but developed according to the framework.

- *Brief Access Technology Scenario*. UMTS in city and urban parts of journey, WLAN connectivity in stations, GSM connectivity everywhere. Digital Terrestrial broadcast, Bluetooth in each carriage.
- *Principal User*. (This is elaborated from a person described in a context perspective earlier). The user/customer is a business executive who is travelling to a meeting. He is male and he is in his 40s. He needs to communicate with his office, with other colleagues in his organisation and with clients. He has access to online resources on his corporate Intranet which he uses in his business. He also submits report and documents. He discusses documents with his colleagues on the phone and works on them at the same time. He is available to his colleagues and clients/ business contacts all the time, but is able to divert calls and messages to a secretary. He has extensive resources to spend on communication and information exchange.
- *Home–work management*. He has a family, but he travels away from home a great deal. He keeps in touch electronically with phone calls every day, images and video messages and the occasional video call. He is also able to check on his young son at the nursery via their webcam. A suggestion that he would also have a lover was vetoed by the project group, despite the interesting perspective on communication management that this involves.
  - He has some friends with whom he shares an enthusiasm for old cars, and other friends he sees on frequent trips to the city.
  - He lives in a suburb and uses his car, trains and planes for travel. Most regional trips are done by train due to bad congestion on the roads.
- *Location: Medium Distance Train*. The user is on a train that travels from the city centre across country for a trip of several hours. The train travels at speeds of 70 kph in urban areas and 130 kph in rural areas. It stops every 15 minutes for three minutes at stations. The business class coach on the train has a power supply for travellers. The stations have WLAN connectivity and the train travels through areas of UMTS connectivity and continuous GSM connection. However, it does go through tunnels

that take up to 15 seconds to pass through. Within the train there may be 100 people doing similar activities to our user. The train also has 200 students going home for the holidays, who are making extensive use of instant messaging, voice calling, on-line gaming and web browsing.

- *The Application Packages*. Network Supported Work Tools *Travellers Aid, Social Networker*. He uses a set of applications that are provided by this employer to link him into the corporate network and to enable communications with clients and business partners. The firm use a customised product provided by a major vendor that integrates multimedia data over a variety of networks. The package enables all the standard business documents to be worked on and shared, access to the corporate Internet and the Internet for communication, information and making travel arrangements, placing secure orders and signing electronic contracts. The system will filter messages and web pages for unnecessary images.
  - He uses an Internet travel company web site on the PDA to book hotels and a taxi.
  - He communicates with his friends and family on a separate integrated e-mail/messaging account through a web interface and the telephone.
- *The Devices*. He has several devices; a small mobile phone, a small PDA and a laptop computer. All the devices are provided by his company and access services provided through a single service provider buying in connectivity from infrastructure companies. He has a large brief case and a jacket pocket to carry them. Devices cannot be used in a moving car while driving, except to take telephone calls, or listen to e-mails and documents being read. The PDA and phone are seen as more convenient to use than the laptop at particular times, especially on crowded trains at rush hour.

9.6.2 Scenario Instances or Use Cases

He will conduct one-to-one and multi-way phone calls, upload and download documents up to 10 Mb and on this occasion a 500 Mb video file for a presentation. For an hour he will discuss a multimedia document with a colleague on the phone while editing the text in real time together. During this time he will check his e-mail three times in response to alerts, downloading five large files to his laptop. He usually checks his e-mail on the PDA, but can synchronise with the laptop and download larger items too. Only one of the downloads is important and must be done immediately, the rest can be done whenever the network connection is fastest and cheapest.

- He watches the business news on a subscription television on demand service.
- He looks at his child in the nursery through a low-quality video link to his laptop web browser.

- He receives some small images from his children on his private messaging service, although routed through the company network.
- He books a taxi and checks a map of the place he is visiting.
- He uses specific data applications: file transfer, e-mail/multimedia messaging, voice phone call, synchronous document sharing, video on demand, broadcast radio, cooperative document editing and web browsing including video stream.

For the testing and design, specific convergence scenarios were developed to quantify the conditions when "simultaneous" use would be put into operation. Most of these relate to demands of individual users as they switch data applications or move into different environments, but some relate to other types of events, such as this example: while the train is in a small town station engineers take down UMTS coverage in the area for five minutes and the system transfers all calls/connection to WLAN, GSM and General Package Radio Service (GPRS) (Table 9.7).

9.7 Building on User Scenarios: Service Product Scenarios

The traditional telecommunications industry has sold bundled applications and services that they guarantee to provide at a particular quality of service level. By contrast, the existing Internet is predicated on flat rate access, with no guarantees on the performance of any applications ("best effort"), which are the responsibility of end users in an end-to-end market. In the future these two models are likely to come together. Mobile Network Operators (MNO) or Mobile Virtual Network Operators (MVNO) will offer network service bundles or Service Products that end users or resellers will be able to use to run a range of end-to-end data applications at a defined quality of service level. The framework can be used to suggest possible Service Products, based on the people, applications and places that will be required. This is used in development and testing of the technology, market studies and assessments of costs, and the likelihood of the role-out of infrastructure and commercial services that will be used to offer that particular product. A Service Product has a socio-technical definition, linking characteristics of particular services, delivered by different bearer services and the requirements to provide a specific marketable Service Product. These products are defined by the particular network services they offer, the locations they are offered in, the price at which they are offered, the quality of service promised and expected, extensions to service available, variations within product (e.g. on pricing, add-ons, etc.) and the market at which they are aimed. Six example Service Products are suggested based on context scenarios. These are not original ideas, but are examples derived using the framework.

Of course this is not an exhaustive list, just some suggestions from the small number of context scenarios. It demonstrates how some generally

Table 9.7 Scenario instance: switching all calls in small town between networks when upgrading network

Specific activity	Data Applications	Network services	Geographical characteristics	Cell types	Network Availability	Mobility	Device	QoS expectancy
250 users	File transfer	Unrestrained data transfer	Outdoor-urban	Micro	UMTS	Vehicular	Phone, laptop, PDA	Medium to high
File sharing	Voice calls		Indoor-urban	Pico	WLAN 802.11a	Stationary		
Phone calls	Web browsing	Speech	Outdoor-suburban	Micro	GSM/GPRS	Walking		
Web browsing		Interactive services				Vehicular – highly vehicular		

accepted Service Products, such as mobile VPN connectivity and simple rich voice, can be found through the scenario method, and also the identification of Service Products by the city dweller and independent business person.

9.7.1 Mobile Internet/Business Intranet

This service can be sold to corporations as part of a "Mobile Enterprise" product, to provide their mobile staff with integration to their company's voice, video and data networks in locations such as airports, offices, on the motorway, trains, etc. Service is provided at a flat rate regardless of international or network roaming and is sold to commercial customers, often as part of a broader service agreement, such as Virtual Private Network (VPN) provision including voice. Users have Network Supported Work tools and use multiple devices (phone, PDA and laptop) which may often be provided or configured by the service company. This is a product within the context of the mobile Internet business model.

9.7.2 Consumer Multimedia Calling

This is a Service Product for people who like to keep in touch, send messages, make low-quality video calls or multimedia calls, send personal photos. Access to information is not important. These users have a version of the Personal Networker Package. This is essentially a simple rich voice product based on extending the functionality of the mobile phone application package and making revenues from occasional use of value-added services.

9.7.3 Multimedia Consumer Service

This is a service based on a mass-market consumer using the Media Portal Package, who principally wants access to media products to be able to have video on-demand, games, download and share music, and video files, etc. For customers with PDAs and PDA+/Web Tablets, in car entertainment, wire-replacement home systems or possibly maxi-phones. This is a product that an operator could provide to a content or entertainment brand reseller, MVNO, or as part of an operator-branded entertainment service.

9.7.4 Emergency Service Communications

This provides radio, data and video communication between emergency service workers and their control bases (fixed and mobile). It requires reserved channels and priority access in certain circumstances. There will be corporate (government) negotiation of terms and a relatively low number of users. It must work in a wide range of locations, but also respect sensitive equipment in operating locations such as hospitals. The emergency services tend to have their own systems, for example, Terrestrial Trunked Radio (TETRA), but multi-bearer services that can simultaneously use public networks or other private systems may become a future requirement.

9.7.5 Independent Business Person's Service

This is a service for people who work from a car, other people's offices and home, and run their own business. They need to be in touch all day and night, send information to clients, conduct marketing and all the usual business activities, while on the move. They use the mobile office version of the Network Supported work tools. This product provides office Internet access, multimedia calling, file transfer, location service, etc. These customers can pay for a premium service, as they may also use it as a wire replacement, and have a high reliance on receiving a high quality of service.

9.7.6 The City Dweller's Service

This service is for the urban dweller, a chatterer or a commuter, who uses the City Survival Kit or Personal Networker. It offers messaging services and voice calling, basic Internet access for access to network applications, such as banking, shopping, travel information, etc., and location services. The service is for urban use, and is not expected to perform in rural areas. This user will probably pass though hotspot areas frequently and require service in offices, public spaces and outside.

9.8 Conclusions and Discussion

9.8.1 Communication and Cooperation

Social scientists have to struggle to define what they can offer in understanding or shaping the development of new technology. This is especially so for those doing user research. Areas of influence in industry include the following: marketing, system configuration, conceptual design and company strategy. In many of these areas commercial consultants, market researchers and internal researchers are the traditional sources of expertise. The FLOWS project has been unusual in that social scientists are working as part of a research and development team that is not creating end-user products, and is indeed not creating anything physical. The social scientists are in a role that is usually taken by engineers who are dipping into user requirements analysis or techno-economic modelling. It is a challenge to find ways to bridge the communication and knowledge gap between research engineers and social scientists.

This project highlighted the significant learning needed to understand the requirements and expectations of the engineers, and the models of users and market that are implicitly used. The scenarios provide the framework for subsequent learning, and although there may not be a great of change in the research programmes, since they are defined in advance, the various technical and social science work programmes can be tied together more closely at the time of dissemination and presentation of results.

9.8.2 Scenarios for Future Wireless

This scenario framework provides ways to tie a range of user research to the requirements of the engineers, but also opens up an agenda for future social science research. It is clear that certain groups of users, places and devices, and uses are very little researched, but will be important. There is a great deal of research on teenage girls and on certain mobile workers, but much less on a cross section of city dwellers, "non-executive" peripatetic workers, or clipboard professionals such as doctors, vets and engineers. Particular areas of research that could be developed in the future include: use of information and broadband on wireless devices; locations of mobility; WLAN hot spots; linking home and public spaces; control of technology and use within particular spaces or among particular groups.

While there is qualitative research on the way people organise their lives, work and relationships, and use ICTs in everyday life, a weakness is in the availability of quantitative data, especially from the operators. Most of the data is rather generic, or gathered from studies of wire-based applications such as within the FLOWS study. Of course it is not possible to have data for service and markets that do not exist, but more detailed figures on the emerging patterns of use of different types of users should be available from operators in some form. Even the research departments of operators have difficulty in obtaining data from the operating division of their own companies because of commercial sensitivity. However, without this data, for example, on the average number of calls or e-mail downloads of business customers according to different locations and times of day, it is hard to build up the scenarios.

Even though it may appear a little mechanistic, this framework is an important example of the work that needs to be done to systematically draw together qualitative and quantitative work on different people and places, devices and applications and present it in a form that can be quickly turned into scenarios for engineering or service design. It can be used to add depth to the scenarios with reference to supporting research and build up a library of examples. The user scenarios can also be used as the basis of new user research to investigate people, such as the travelling executive or family on holiday and the conditions under which they may turn to technology at different points in their lives.

Much scenario work is conducted by engineers, with little knowledge of the large amount of research on use of ICTs, or of the methods and approaches developed to understand the relationship between technology and people. This chapter hopes to provide some useful inputs for those within the industry to incorporate into their work.

Acknowledgements

Scenarios in this chapter were developed jointly with Lisa Lee at the University of Edinburgh.

References

3GPP (2003) Services and service capabilities: 3GPP TS 22.105 V6.2.0 (2003–2006), 3rd Generation Partnership Project; Technical Specification Group Services and System Aspects Service aspects; Services and Service Capabilities (Release 6). 3rd Generation Partnership Project, Sophia Antipolis.

Aguiar J, Correia L, Gil J, Noll J, Karlsen ME, Svaet S (2002) Definition of Scenarios (D1). FLOWS-IST Deliverable, D1, IST-TUL/ FLOWS Project/EC IST Programme, Lisbon.

Beyond 3G Project Cluster (2002) *A Vision on Systems Beyond 3G. Beyond 3G Project Cluster*, European Commission IST Programme, Brussels.

Brown B, Green N, Harper R (eds). (2001) *Wireless World: Social and Interactional Aspects of the Mobile Age*. Springer, London.

Cook K (2002) Beyond 3G: What Will the Users Want? Paper presented to Wireless World Research Forum Working Group 1, 2002. Wireless World Research Forum.

Cook K, Aftelak A (2001) User Driven Intelligent Service Automation. Paper presented to Wireless World Research Forum.

Crisp MJ, Foley S, Lievonen M (2002) Location Awareness in Working and Domestic Life. Paper presented to The Social Shaping of Mobile Futures: Location! Location!, Third Wireless World Conference, University of Surrey.

de Gournay C, Mercier PA (1998) Entre la vie privée et le travail: décloisonnement et nouveaux partages. Paper presented to Penser des Usages, Bordeaux, July 1998.

Dietrich NA, Eichner U (2001) Wireless Enabled Networked Communities. Paper presented to Wireless World Research Forum Kick Off Meeting, 6–7 March 2001.

Dix A, Beale R (1996a) Information requirements of distributed workers. Remote cooperation. In: Dix A, Beale R (eds). *CSCW Issues for Mobile and Teleworkers*. Springer-Verlag, London; pp. 113–144.

Dix A, Beale R (eds). (1996b) *Remote Cooperation. CSCW Issues for Mobile and Teleworkers*. Computer Supported Cooperative Work. Springer-Verlag, London.

Fortunati L (2002) Italy: stereotypes, true and false. In: Katz JE, Aakhus M (eds). *Perpetual Contact: Mobile Communication, Private Talk, Public Performance*. Cambridge University Press, Cambridge.

Grant D, Kiesler S (2001) Blurring the boundaries: cell phones, mobility and the line between work and personal life. In: Brown B, Green N, Harper R (eds). *Wireless World, Social and Interactional Aspects of the Mobile Age*. Springer, London.

Haddon L (2000) The Social Consequences of Mobile Telephony: Framing Questions. "Sosiale Konsekvenser av Mobiltelefoni", Telenor, Oslo, 16 June 2000.

Haddon L (2001) Domestication and Mobile Telephony. Paper presented to Machines that Become Us, Rutgers University, New Brunswick, New Jersey, US, 18–19 April 2001.

Haddon L, de Gournay C, Lohan M, Ostlund B, Palombini I, Kilegran M (2001) From Mobile to Mobility: the Consumption of ICTs and Mobility in Everyday Life. Report of the Cost 269 Mobility Group, Brussels.

Hickey S, Pulli P (2001) Developing a User Centric Multisphere Model for Mobile Users. Wireless World Research Forum, Workshop submission.

Katz J, Aakhus M (2002) *Perpetual Contact: Mobile Communication, Private Talk, Public Performance.* Cambridge University Press, Cambridge.

Kopomaa T (1999) Speaking mobile: intensified everyday life, condensed city – observations on the meaning and public use of mobile phones in Helsinki. *Cities in the Global Information Society Conference.* Newcastle upon Tyne, UK, 1999.

Laurier E (2001) The region as socio-technical accomplishment of mobile workers. In: Brown B, Green N, Harper R (eds). *Wireless World: Social and Interactional Aspects of the Mobile Age.* Springer, London.

Lehr W, McKnight LW (2002) *Wireless Internet Access: 3G vs. WiFi? ITS Conference.* Madrid, September 2002. Conference Paper \lt;http://itc.mit.edu/\gt;.

Lie M, Sorensen KH (1997) Making technology our own? domesticating technology into everyday life. In: Lie M, Sorensen KH (eds). *Making Technology Our Own? Domesticating Technology into Everyday Life.* Scandinavian University Press, Oslo.

Ling R (1999) I am Happiest by Having the Best: The Adoption and Rejection of Mobile Telephony. FoU Rapport 15/99, Telenor Forskning og Utvikling, Kjeller.

Ling R (2004) *The Mobile Connection: The Cell Phone's Impact on Society.* Morgan Kaufmann Series in Interactive Technologies. Morgan Kaufmann, San Francisco.

Ling R, Yttri B (2002) Hyper-coordination via mobile phones in Norway. In: Katz JE, Aakhus M (eds). *Perpetual Contact: Mobile Communication, Private Talk, Public Performance.* Cambridge University Press, Cambridge.

Nippert-Eng CE (1995) *Home and Work.* University of Chicago Press, Chicago.

Sherry J, Salvador T (2002) Running and grimacing: the struggle for balance in mobile work. In: Brown B, Green N, Harper R (eds). *Wireless World: Social and Interactional Aspects of the Mobile Age.* Springer, London.

Stewart J, Dorfer W, Pitt L, Eskedal T, Gaarder K, Winskel M, Evans D, Williams R, Stimming C (2003) Cost and benefit of use scenarios: the selection environment for MIMO-enabled multi-standard wireless devices including cost benefit analysis of various convergence technologies, FLOWS Deliverable D12. University of Edinburgh/FLOWS/European Commission IST Programme, Edinburgh.

Stewart J, Pitt L, Winskel M, Williams R, Graham I, Aguiar J, Correia LM, Hunt B, Moulsley T, Paint F, Svaet S, Michael B, Burr A, Eskedal TG, Yin V, Stimming C (2002) FLOWS Scenarios and Definition of Services. FLOWS-IST Project Deliverable, D06. University of Edinburgh/FLOWS/ European Commission IST Programme, Edinburgh.

Thank DV, Vanem E, Tran DV (2001) Towards user centric communications with the virtual terminal. *Teletronikk,* **97**(1): 106–126.

Townsend AM (2002) Mobile communications in the twenty first century city. In: Brown B, Green N, Harper R (eds). *Wireless World: Social and Interactional Aspects of the Mobile Age.* Springer, London.

UMTS Forum (2000a) Report on the Extended Vision of UMTS. 10, UMTS Forum.

UMTS Forum (2000b) UMTS Third generation market – Structuring the Service Revenue Opportunities. UMTS Forum, London.

UMTS Forum (2002) IMS Service Vision for 3G Markets. Report from the UMTS Forum, 20, UMTS Forum.

Velez FJ, Correia LM (2000) Deployment Scenarios for Mobile Broadband Communications. Paper presented to *Proc. of PIMRC'2000 – IEEE 11th International Symposium on Personal, Indoor and Mobile Radio Communications.* London, UK, September 2000.

Velez FJ, Correia LM (2001) Impact of Mobility in Mobile Broadband Systems Multi-service Traffic. Paper presented to *Proc. of PIMRC'2001 – IEEE 12th International Symposium on Personal, Indoor and Mobile Radio Communications.* San Diego, CA, USA, September 2001.

Williams R, Edge D (1996) The social shaping of technology. *Research Policy,* **25**: 856–899.

Williams R, Stewart J, Slack R (2005) *Experimenting with Information and Communication Technologies: Social Learning in Technological Innovation.* Edward Elgar, Cheltenham.

Woodward B, Istepanian RSH, Richards CI (2001) Design of a telemedicine system using a mobile telephone. *IEEE Transactions on Information Technology in Biomedicine,* **5**(1): 13–15.

Instant Messaging and Presence Services: Mobile Future, CSCW and Ethnography

Philippe Rouchy

10.1 Introduction

The aim of this chapter is to explore Instant Messaging and Presence Services (IMPS), developed under the technical–commercial umbrella of Unified Messaging Services (UMS) with the help of ethnographical studies of mobile phone users. The task is not as straightforward as it seems. Although there is widespread use of ethnographical investigations in system design and mobile systems, specifically in a work environment (Weilenmann *et al.* 2002), it seems that difficulties continually prevent researchers from informing the design of those systems. Also, because mobile systems are used outside the office environment, academic studies of this technology have to rely on channels of information other than the regular project development usually available to academic researchers.

While the application of ethnographical methods to large information systems suggests selected reservations about careless application of design guidelines, the relative force of those studies does not seem to diminish when dealing with the "information rich" contexts in which people use mobile technologies. Many researchers have employed various participants' studies, for example diary studies (Palen and Salzman, 2002), in seeking to discover users' engagement with specific technologies. The problem of ethnography, as we see it, has rarely been a question of observation *per se*, but rather a clear articulation of the level of analysis. The aim of ethnographical studies is to achieve a deeper understanding of computer-mediated communication, distributed information systems and discrete technologies, such as mobile systems. One of the first pitfalls ethnography helps to avoid is the superficial understanding of usability. UMS presents a perfect challenge to the notion of usability. UMS encompasses technical developments in telecommunications by operators, and commercialisation or marketing of mobile devices through specific services, such as IMPS. This technical–commercial context suggests dealing with users of new-generation mobile phones by considering the notion of usefulness rather than of usability.

This chapter shows why evaluative ethnographical studies are suitable for studies of usefulness of mobile systems. Some researchers have demonstrated the importance of full-scale sociological competence (Button, 2000) when dealing with information navigation on the web in institutions (Harper, 2002) or with users of mobile systems at the level of rules of behaviour (Murtagh, 2002). The articulation of analysis at both institutional and personal levels is especially necessary when dealing with IMPS development because the technical–commercial levels operate as an invisible hand over users' experience. It is necessary to grasp three points: the UMS's technical–commercial environment of such system development; its implication for the user as a simple agent of commercial transaction with the service provider; and the difference it makes for experienced users of mobile phones. In light of these issues, technical–commercial information concerning UMS is reviewed and we consider how an evaluative type of ethnography can use IMPS's technical–commercial concepts (presence, instant messaging, groups and shared content) as a guide for looking at existing practices with mobile phones. Those concepts will guide the line of investigation through vignettes (Orr, 1996) of mobile phone use. These reported scenes are then compared with concepts developed in the mobile marketing domain.

10.2 UMS: Technical–Commercial Future of Mobile Systems

The point of departure is a sociological analysis of the relevant fields (marketing, corporate and technical collaboration) involved in the future of mobile technology. This entails an exploration of the mobile system industry in light of its own industrial initiatives, commercial endeavours and cooperative technological developments. To start with, UMS is a technical–commercial initiative that does not translate directly into a commercial product sold to customers. It is mainly a business agreement (Woods, 2002a) that defines how mobile operators will offer a set of standards for mobile communication. These standards are visible to the public only indirectly, via the operators' marketing campaigns. Marketing mediation partly explains why UMS is not a concept widely known to the public. It is not sold as a comprehensive set of services, but as a selection of services. In a commercial context, marketing studies inform business decisions about what services and tools should be developed to achieve commercial success; that is to maintain or increase the products' sales. The case of UMS illustrates how the commercial approach is an integrated and inseparable part of the way in which decisions are made to develop and launch integrated technologies of the next generation of mobile phones and services. In other words, commerce is a good enough reason to start to develop new technology and new services. However, for clarity of presentation, the commercial aspects of UMS are separated from its technical aspects.

10.3 Open Industry Initiatives: The Commercial Approach

In 2001 Ericsson, Motorola and Nokia decided to collaborate on instant messaging systems for mobile phones. Their objective was to promote mobile services in the domain of messaging. This is partly done under the assumption that notifying users on what kind of device – personal computer (PC), personal digital assistant (PDA), mobile telephone – they can reach their interlocutors will be a significant asset in future communication forms. From the companies' perspective, this industrial objective should translate into tools available in the third generation (3G) of mobile phones and other mobile devices (such as PDAs and laptops). The strategic choice of those companies is to keep the door open to other companies to join exponential services development. The justification for this "open industry" initiative is both commercial and technological. The open approach reduces the cost of research and development of similar but competing products. It avoids standardisation battles by not providing the opportunity for monopolising software development. Although the actual design of the software is left to the developers of each company, the open industrial approach seeks to address another parallel issue in mobile telephony, namely establishing telecommunication standards.

Successful commerce is a core incentive for the open industry initiative. The commercial interest concerning users of 3G mobile phone technology translates into the availability of communication services. These should not be proprietary to the devices that customers buy. Essentially, anyone having a newer mobile phone of any given brand should be able to send any kind of message – e-mail, text message (Short Messaging Service, SMS) multimedia message services (MMS) – to other owners of mobile or stationary devices, or to chat live to them. The technological incentive follows closely from the commercial one. If the idea is to get more and more people using as diverse a range of telecommunication services as possible, the technological issue is straightforward. The open industry initiative offers technology that is not proprietary. At this level, project management is the main concern of the consortium of companies. They have to develop the technology together and to share their progress, and then create and market a meaningful package of telecommunication and software systems for customers. The development of UMS emerges around cooperative initiatives, such as the Open Mobile Alliance (OMA). Such initiatives make software and telecommunication standards compatible in order to facilitate future technological development in a competitive market. UMS is an umbrella covering a set of integrated services that operators want to sell to customers. The idea is to start with a mutual agreement whereby the parties take part in the joint development of products. The model for a simple industrial agreement over the development of a new legal structure comes from other industrial initiatives, certainly within software development (open software initiative) but also in the telecommunication industry (e.g. Bluetooth and SyncML). From the sociological point of view of this study, UMS is already an integrated product of technology development in software industries, telecommunication networks and marketing.

At the forefront of marketing concepts informing the technological development of mobile services is IMPS. The aim is to work out ways to integrate existing commercially and technically successful products in four areas:

1. Notification devices indicating the status of a communication (called "presence").
2. Instant messaging (on the model of SMS) using communication devices other than two phones (e.g. phone to computer).
3. The composition of lists of receivers (called "groups"), which work more like Internet chat rooms than e-mail lists.
4. The use of accounts where users can share information (called "shared content").

The mobile system technology development starts with a non-proprietary agreement concerning instant messaging. The industrial agreement on the telecommunication front does not prevent the companies from developing differentiated (and potentially competitive) software strategies. The main commercial incentive for those companies is to attract new customers and upgrade existing customers' services to the 3G networks. The ultimate commercial success would be for the customer to financially support and be an active part of their network investment when leveraging a second-generation (2G) network to 2.5 and 3.

10.4 UMS: The Technological Integration Approach

From the technological point of view, UMS requires that instant messaging solutions and identification of the sender and caller are fully interoperating solutions. It means that the solution should:

1. work with both wireless and wired networks interchangeably;
2. support a variety of transport protocols (as telecommunication protocols are developed often independently of each other by independent interest groups throughout the world);

3. support integrated software packages, such as instant messaging, chat applications, access to shared content and information, on the status of the current connection;

4. enable the transcription of data by any devices, regardless of their type (e.g. PDA, mobile phone, laptop and stationary computers);

5. increase the number of sources of information that a mobile device can access;

6. take advantage of other communication networks, such as the existing Internet and its related software technologies including web design and multimedia content.

The telecommunication companies (network operators and hardware companies) design their technical specifications of interoperability between platforms within the telecommunication area (Woods, 2002b). This means that the industrial consortium develops software and interoperability solutions from scratch (which needs testing) (Perez, 2002). The agreement driving the actors of the development of integrated mobile systems does not deal with the definition of software. The software is not the crux of the commercial matter. There is no prior and specific agreement on how data should be presented. In light of this, it is not surprising that the commercial services will be primarily handled by telecommunication operators.

10.5 Server Infrastructure for Developing UMS

Due to the freedom regarding the development of interfaces and software languages, the consortium of people working on UMS deals with web software, such as PHP Hypertext Pre-processor (PHP), which offers connection between web-based data and communication protocols like Wireless Application Protocol (WAP) rather than data architecture *per se*. Their interest concentrates mainly on accessing data and on a server protocol that can be differentiated in three domains of development:

1. The telecommunication consortium's main interest is putting together a common protocol. To reach such a goal, they need to share their work on a dedicated server. However, they must also have an agreement on how each actor accesses the server and interacts with the others.

2. In respect of the mobile devices, they have to decide on a client–server protocol that allows mobile or desktop access to providers of UMS.

3. To ensure interoperability between the development server and existing wireless domains and Internet-based providers of UMS, they produce a server–server protocol, which works as a gateway to other domains.

10.6 Interoperability Protocol in Telecommunication

In the telecommunication industry's technological development, deciding on a protocol is necessary as it is the way information is shared between sender and receiver devices. For example, a consortium of interested parties can choose a standard developed by a non-profit organisation of professionals organised as a technical and scientific community. In mobile telephony and services, technologists work with the Institute of Electrical and Electronics Engineers (IEEE) and the Internet Engineering Task Force (IETF). Non-profit technical professional associations such as these propose standards (e.g. see http://standards.ieee.org/) for electronics and telecommunication. In the case of the technological development of mobile telephone standards, as developed under the UMS umbrella, the consortium defines guidelines such as:

1. to comply with existing industrial standards (available through those professional associations);
2. to include existing communication technologies (e.g. SMS, MMS, General Packet Radio Service (GPRS), and Transmission Control Protocol/Internet Protocol (TCP/IP));
3. to work with other standardisation bodies (e.g. the WAP Forum, 3G Partnership Project (3GPP) and IETF);
4. to create standards for their services (e.g. Common Profile for Instant Messaging (CPIM), Multipurpose Internet Mail Extensions (MIME) and Instant Messaging and Presence Protocol (IMPP).

The engineers' labour force is organised on the model of open communities of specialists. They work together by belonging to mailing lists and working groups, which define their goals and milestones. Then they publish drafts on the Internet and request comments from the technical community. It is common practice in such communities to publish ideas in working towards a common goal. Most of the issues emerge from such discussions (Arief *et al.*, 2002). In this case, the companies' initiative makes every effort to bring people to work together for a common standard. This is done to allow service providers to have their own end-user community. In return, it obliges the industry-leading coalition to be open to participation beyond the group of project originators.

10.7 Evaluative Ethnography in CSCW

To illuminate the evaluative dimension of ethnographical work, it is necessary to recall how pioneering works in Computer-Supported Cooperative Work (CSCW) proceeded with system development. Ethnographical studies of air traffic controllers (Hughes *et al.*, 1992), bank clerks and managers (Randall *et al.*, 1994) expert systems in lawyers' firms (Blomberg *et al.*, 1996; Whalen and Vinkhuysen, 2000), large service companies and international financial institutions (Harper, 2000; O'Hara *et al.*, 2002) have offered invaluable information on what to expect of users of systems by providing relevant dimensions of their work practices. Ethnography cannot hold strong claims in engineering terms: it is neither a method nor a complement of existing engineering practices. When thinking about the way ethnography can be put into profitable use for knowing the studied phenomenon, it is necessary to understand ethnography as a basis for an empirical investigation of work. It is not the point here to get into the crucial difference between ethnography in social science and ethnography for system design. However, to get a fair understanding of the latter in the context of UMS development, it is necessary to recall the difference between the relevance of ethnography for CSCW system development and ethnography for mobile technology.

In an essential distinction offered by ethnographers of system design (Hughes *et al.*, 1994), four uses of ethnography for system development can be distinguished:

1. Concurrent ethnography: where design is influenced by an on-going ethnographical study taking place at the same time as the systems development.
2. Quick and dirty ethnography: where brief ethnographical studies are undertaken to provide a general but informed sense of the setting for designers.
3. Evaluative ethnography: where an ethnographical study is undertaken to verify or validate a set of already formulated design decisions.
4. Re-examination of previous studies: to inform initial design thinking.

Within CSCW, studies of work and particularly ethnographical studies of work, have mostly been done to inform system designers during system development. It is well known that systems fail because their designers pay little attention to the social context or circumstances of work. In this situation ethnography can play an informative (but not corrective) role in respect of the inadequacy of existing methods for requirements elicitation and work analysis.

In CSCW, the plausibility of diagnosis through ethnography came from a mutually informed way of working between system developers and professional ethnographers. In the air traffic control study (Hughes *et al.*, 1992), it is made clear that cooperation between professionals took place under the auspices of the technological and organisational conditions in which the system developers were working. The transmission of ethnographical input (about the manipulation of "flight strips" for the management of air traffic) fed back to the system designers to inform them about the reality of the controllers' work and its coherence with the technical development of a new system. The practical instantiation of the cooperation between ethnographers and system developers was the debriefing meetings. Information went both ways. System developers gave indications to the ethnographer for specific things to look at. Given the literary nature of ethnographical materials, system developers spent the time and attention necessary to read the ethnographical research results.

10.8 The Comparative Method

It should be noted that the air traffic control study was intended for large-scale system design and cooperative work settings. We suggest that there is much to learn from their distinctions for importing ethnographical insights into the UMS context. First, it quickly becomes apparent that evaluative ethnography is the most suitable approach for the study of system development within UMS. Some justification is in order:

1. The development of a system like IMPS within the UMS area is dictated by the development of technological integration as well as commercial breakthroughs. The nature of mobile technology development deals with a highly standardised industry where the technology is available to users all over the world. In contrast with most system design in the workplace, IMPS technology is not designed for specific work practices (e.g. traffic control, office document management and technical expert systems). For example, one obvious use of ethnographical insight for system design comes from the necessary updating of computer-distributed systems that was put into place in the 1980s. Upgrading and installation of multimedia facilities in those systems have been, or are being, achieved. An example is the Swedish emergency hotline, which is using ethnographical studies to get a careful assessment of what should be retained and what should be improved as a result of updating their computer-aided dispatch (CAD) system (Pettersson and Rouchy, 2002). In the IMPS business, the design of interoperation between systems constitutes an offering on the lucrative market of mobile systems, private and corporate.

2. As seen before, the development of a system such as IMPS starts with corporate agreement about the types, range and number of telecommunication standards to employ. As such the user has no part in the development of this telecommunication system. It is therefore impossible to use ethnography for designing a system whose existence is the fruit of a corporate agreement. Guidelines and design solutions for developing IMPS systems are on their way.

3. For the IMPS system, ethnography cannot evaluate the design solutions of the network system because they are in the domain of engineering

conventions (standards). In the domain of standard setting, interest groups have to create the opportunities for agreement. In the case of the IMPS system, ethnography has nothing to offer in the evaluation of standards, nor in the details of the business agreement. Nevertheless, ethnography can evaluate current communication practices. As such it can provide an assessment of the IMPS system at the users' end by offering a comparison between the actual use of mobile devices and what UMS's approach expects from users of the IMPS system. Two related arguments advocate the rationale for ethnographical studies of IMPS systems in a prospective–evaluative way. Firstly, the top-down approach of the mobile business risks overestimating the number of consumers for services. Secondly, because knowledge concerning the user is prospectively driven by commercial and marketing considerations, ethnography represents an alternative approach for the same task. It underlines people's practices within the relevant domain of activities' showing their choices, expectations and routines.

10.9 CSCW, Information Technology and Vignettes of Work

Ethnographical studies of users' need and practices are post-developmental, suitable for assessment of, or for developmental insights about, technology in its working environment (Hughes · et al., 1994). IMPS system development is based on guesswork about what future customers of mobile services might do. By performing ethnographical studies, the viewpoints are shifted from UMS's consortium approach to users' experience. CSCW studies reflect a growing interest in low-cost technology, such as personal devices (e.g. mobile phones, PDAs, laptops). Personal devices are designed as consumer objects related to people's personal lives as well as work environments. One of the difficulties of studying mobile phone users has been the need for researchers to report on highly varied social contexts in which the technology is used. As operating such devices is easy, considerations about usability are often pointless if not related to consideration of the usefulness of the proposed services. Julian Orr's ethnographical vignettes of work (Orr, 1996) facilitate the study of the user of personal technology in relation to consumption of telecommunication services. Such reported scenes are used to capture aspects of social interactions in short but detailed ways and they can cover a set of activities by different people in different contexts. Such vignettes offer the possibility of comparing the IMPS concepts (presence, instant messaging, group list and shared content) with four different situations where people use their mobile phones; forwarding information in an office, a mother's remote access to her child, a teenager's use of the phone book and calendar-shared information.

10.9.1 Vignette 1: Presence and Reachability

According to IMPS's technical–commercial approach, "presence" is a function of future mobile devices accounting for the status of communication. For example, callers will be able to check if the receiver's device is available to receive messages. In the work environment, people will be able to get information about the availability of their interlocutor (in a meeting, at lunch, client visit, holiday, back in 10 minutes, etc.). This ethnographical vignette deals with the issue of parenting rather than work environment in a situation where parental control over a child is a constitutive part of the relationship. Here is the case of a single mother dealing with her child.

Most days of the week, a single mother with a 11-year-old son works at home using her computer which has a dialup connection to the Internet. She also has access to a fast Internet connection in her office at the University. Her son finishes school at 2 pm and comes home, where she prepares a meal for him. The boy spends his spare time between several activities, including playing ball with friends in the schoolyard. If the children have agreed to play together, the boy needs his mother's permission to play outside for a given period of time. He borrows his mother's mobile phone so she can call him from home when necessary. It also allows the child some flexibility, since he can negotiate his schedule with his mother. He often reschedules his activities to play computer games with his friend from around 4 pm to 6 pm. These activities are seasonal as it gets dark about 4 pm in winter. Children do not spend so much time outside then and their parents have to organise their afternoon in other ways.

The interoperability of a system in a real social condition has been insufficiently studied. One of the obvious conditions for interoperability of systems in everyday life requires people possessing multiple systems, such as mobile phones and a connection to Internet at home, or at the office, working together. The vignette shows that there is felicitous condition for using interoperability in everyday life. The relationship between a child and his mother illustrates the importance that trust, negotiation and expectations play. These are, of course, different from those in a relationship between colleagues in an office. The IMPS's presence function is essentially a function of notification: something that gives a time frame and an indication of conversational expectations. The IMPS presence function is very much a translation of the office phone functions. In an office context, expectations of the reachability of people are regulated by office hours, meetings and other activities, as well as status within the organisation, etc. The IMPS presence function is very much built upon the issue of reachability in conversational contexts. Its use in the private sphere is relevant where stable expectations of reachability exist (such as in this case). IMPS's presence does not offer a significantly extra value in terms of service compared to using a phone line, even during web surfing, to communicate important information, such as time schedules.

The issue of reachability with a mobile phone covers commercial or managerial domains of work where constant monitoring of activities (such as consultancy, management work on several sites) is at stake. In the private sphere, the usefulness is less obvious. This case is an example where the application of the IMPS presence function could be useful. Some may consider this use to monitor another's activities as creating interpersonal disturbances. As presence translates in practice into checking up, some people will use it for exercising misplaced social control. In practice, indication such as "in a meeting" invites other forms of checking (through fixed line phone, e-mail, asking colleagues, etc.). Although the context provides for the understanding of the notification (where "in a meeting" may mean "I am not here" or "I am not available to you") presence is the direct translation of office phone function into mobile communication devices. The critical point about the technical–commercial use of presence is the important role given to the notification of absence. It provides at best an answer, and at worst, a justification for the unsuccessful reachability of an interlocutor.

10.9.2 Vignette 2: Instant Messaging, Relevance and Self-Selection of Information

According to IMPS's technical–commercial approach, instant messaging offers a similar coverage to existing technology in mobile devices. For example, e-mail and chat lists should complement the existing SMS and MMS.

This ethnographical vignette reports how people signify to each other relevant information through e-mails in the work environment of their office. In the example of e-mail exchange of information taking place in an academic department it is possible to distinguish between notifying somebody of something and forwarding information. During the coffee break, Swedish and French academics working in the same department talk about a research seminar taking place in Stockholm. The Swedish academic has thus already identified the French academic as somebody who may be interested in the message. The message is a lengthy description of the seminar in Swedish only:

From: sender@edu.se
To: receiver@edu.se
Cc:
Subject: FW: Topic of the seminar in Swedish

-------- Original Message --------
From: miss x
Sent: Tuesday, April 08, 2003 4:16 PM
Subject: Topic of the seminar in Swedish

The academic institute of … and the research funding body will hold a seminar on 16th May:

- The topic
- A brief explanation of the topic
- A presentation of the two speakers at the seminar
- A detailed programme — a schedule
- A notice saying that the seminar will be held in English
- When it will take place: Friday 16 May, 2003, between 09.00–12.30
- Where it will take place: conference centre, street name, area and city
- Registration: deadline for registration
- How much will it cost: …

For further information and registration: Name (miss x), address and contact information.

Regards
Miss X

Miss X
Institute of …
Address
Contact info

The case of the e-mail communication brings to the fore the issue of relevance and self-selection of messages. In the e-mail context, the issue of reachability is reserved to special cases such as intercultural communication (where translation is involved) and specific institutional or personal contexts (status or private) where people dispatch information to others. The interoperability of instant messaging generalised to the e-mail invites two remarks. One is the issue related to spam e-mails and their availability on a mobile phone. The other is the advantage that users will make of having e-mail or chat messaging on their mobile phone. Keeping in mind that duties between relatives, professionals or groups of peers render communication necessary, it is clear that the IMPS's instant messaging function enlarges the spectrum of messages available on the phone to resemble more and more that of a PC. From the users' point of view, the phone is facing a similar issue that PC communication faces, namely the increase of messages coming from remote communication networks. With a stationary phone, the answering machine can be used for filtering calls. Contact folders into which your e-mail is sorted may become one way to classify your incoming messages. The issue of callers' hierarchy precipitates the issue of over-consumption of communication means. It is the case that users are already customers of a set of communication tools. Although the availability of instant messages from different sources (e-mail, chat room, SMS or MMS) is in a technical integration phase, it shows that operators are interested in new customers willing to purchase a new generation of phones where multioperations are an argument for getting a simplified subscription. From the operator's point of view, the question is not to know if customers are willing to improve the function of their phone. In fact, the economic battle consists of the operator's need to attract a new mass of customers. It is thus crucial for operators to attract the new generations: teenagers. This naturalist approach of marketing is powerful precisely because teenagers themselves (and an increasing part of the adult population) do not understand product cycles and fail to see it as a source of revenue (and thus be able to resist the marketing ruse). The next vignette considers what a teenager does with her mobile phone in public places.

10.9.3 Vignette 3: Group Lists, Constituting or Consulting Them

According to IMPS's technical–commercial approach, the composition of group lists (list of interlocutors) functions like Internet chat rooms rather than e-mail lists. The idea is that people can enter into a discussion group through their mobile phone.

Looking at the constitution of group lists on the Internet, it is clear that they function on a very normative definition of the situations of exchange. For example, groups of gamers, musicians, software developers, dating agency clients and teenage hobbyists have found in the Internet a well-structured channel of exchange. Lists are never constituted *ad hoc* but around the definition of a situation that provides for the identification of common interest and reason of exchange. Studies of Internet chat rooms in computer-mediated communication have shown (Have, 2000) that part of the chat is dedicated to identification through room names, nicknames, to evaluate how newcomers comply to the rules of the community and to reply to well-identified partners of exchange. The issue becomes the context in which people will get involved in those lists. With a stationary computer, people set up their time schedule to dedicate themselves to the chat room of their choice. With a mobile phone, the question arises as to whether access to a chat room list or users' lists may be done on the spot, in the street, or in a public space. The third vignette is, in some sense, a case of social experimentation that takes place in the margins of a well-established framework of communication (a list of well-known callers) and the notion of *impromptu* exchange done on the spot with anyone (potential new callers). As the case shows, this margin is explored by a teenager trying to get in contact with somebody in a supermarket using her mobile phone as a device triggering conversational contact.

Examples of this sort are difficult to observe because of the *impromptu* nature of the situation. It is nevertheless interesting to show how teenagers handle them as they display their own confusion on how to invite contact with unknown people, counting on what they know about lists of callers in their mobile phone. This is how it went.

A man is shopping in the late evening in a local supermarket. He notices a female teenager in the aisle busy looking around rather than shopping. He thinks she may be shoplifting, but he carries on with his own shopping. Later he goes to the only open checkout, where the same teenager is queuing in front of him. She suddenly decides, without apparent reason, to step out of the queue and indicates that the man should take her place. Surprised, he does so. She stands next to him, her back to the queue, using her mobile phone. It was clear that her phone was not ringing, or vibrating. No call was answered as she did not bring her phone to her ear. She could have received an SMS. But reading an SMS would not have justified leaving the queue. The orientation of her body and the use of the mobile phone showed the type of device she used, a 2G mobile phone. It was also possible to discern the navigation menu she was using. She opened an empty field such as "call contact" you find in both the "phone book" and "message" options. She returned to the queue after the man when he had reached the cashier. No conversation took place between them. Both carried on their own business separately as if nothing had happened.

A great deal of interpretation can be placed on those observations, and such interpretations and inferences have some deductive value concerning what people can do with their mobile phone. One thing suggested here is that the teenager using her mobile phone assumes that people know what she is doing with it, namely, consulting the contact list by navigating the menu. As suggested with the example of the stationary computer, the use and constitution of a contact list is a very structured activity performed by communities of members who have to demonstrate the right characteristics to qualify to belong to a community. It reminds us that a great deal of work must be done prior to one becoming a full participant in the community of interest. The demonstrative way in which the teenager created an entry on her mobile phone is a reminder that list and community cannot be constituted without the rules that specify the structured way to join a community. The teenager example shows that mobile phone group lists and users' lists cannot be a shortcut for engaging even minimally into social interaction and controlling inference about the status of the relationship before adding a new entry in her phone list. If the teenager had spoken to the man next to her in the queue and asked for his number, this would have presented an entirely different situation. People will not constitute group lists or users' lists on the spot without prior knowledge of who may constitute the list. On the contrary, those group lists depend on a well-established constitution of pre-existing networks. It also seems unrealistic that people will use their mobile to participate fully in their chat community. The teenager's behaviour with her mobile phone may have been an attempt to get in contact with the person in the queue, offering the mobile phone as a way to suggest an *impromptu* conversation. If this was indeed the case, it proves that group lists cannot be simply the result of superficial engagement in social contacts.

In conclusion, the IMPS's group list function will necessarily be overlaid on existing practices of exchange and cooperation (electronic or otherwise) constituting the group. Again, ethnographical studies of users show the difference between constituting and consulting a list (an entry in a phone book or a user's entry in a group list). In the case of a group list on the mobile, the consulting operation seems the most plausible. From the operator's point of view, the consultation of services is what the operator wants the user to do. The operator's economic motivation for the development of such an integrated new generation of mobile phones requires considering IMPS as a new package of solutions, inviting customers to buy a new generation of phone/handheld computer and their simplified procedure of operator's subscription. The question of pricing those services cannot be considered service-by-service (as the analytic procedure of the vignettes does here) but as a new marketing package offering a new concept for the new generation of mobile phones.

10.9.4 Vignette 4: Shared Content and Information Repository

According to IMPS's technical–commercial approach, the shared content function allows the use of accounts accessing a shared information repository. The idea of this technology is based on today's professional web access to shared content such as databases. It is possible to create other kinds of domains of content, such as directories collectively managed (e.g. open source development pages), or collective annotation systems in project work, related discussion systems or joint editing systems.

People in any kind of organisation work routinely with shared content either with traditional file systems or with online content (Trigg *et al.*, 1999). One of the noticeable features of online-shared content is the way in which users must rely on very clear (step-by-step) operating procedures, generally provided by the organisation. The university example has shown how people using shared information such as calendar and scheduling online could access them on their mobile device, thanks to WAP. This example of the use of the repository function on the mobile phone is particularly enlightening because it is based on the former generation of mobile phones, which support WAP but do not incorporate UMS's functions, as a way of thinking about the new generation of mobile phones. The vignette deals with use of a calendar of course schedules, accessible to all workers in an academic institution.

The calendar, an example of shared content, reveals interesting annoyances when used with older generations of mobile phone. The mobile phone used has MMS, WAP functions and a black-and-white screen sufficiently large for the information displayed. The concern in this section is the problem of navigation via the telecommunication network to access the calendar (as example of a shared content repository). Consider the work that needs to be done for somebody to consult his calendar using a WAP network. After selecting the calendar function on the mobile interface, the user gets access to the server that will connect it to where the calendar is stored. To use such a communication service, the user is obliged to enter a password. Writing the password on a mobile phone is difficult. Practice shows that a short password, of three letters maximum, is the best. Function switching between capitals and lower case letters helps the process of typing. The user needs to choose between services such as a calendar, an address book and a to-do list. When accessing the calendar, the user gets today's schedule comprising basic information such as time, place and activity. The user may check the previous and following days' schedules. Other options are available, such as notes and specification of topic. The detour via a telecommunication network to consult a calendar constitutes an overhead in office practices. The activity of taking notes is performed more easily with a pen in a paper diary rather than with either a stationary computer (implying the repository be located in the office) or a mobile device (implying a telecommunication procedure).

The vignette shows that main annoyances in using the calendar as shared content are not related to the telecommunication system. Observation shows that, with an old-generation mobile phone, there are problems of navigation in accessing services. Shared content works mainly with people consulting rather than creating, modifying and working extensively on the document content wirelessly via a mobile phone. In an office work environment, the shared content becomes interesting if several people involved in a common project are able to modify a particular document. The generality of shared content does not presume what kind of work is carried out on a shared repository. It seems that extensive work on documents will be improbable or at least very limited. The problem is well documented by the difference between the Microsoft Word application on PCs and the version available for Pocket PC (a PDA). Even if this PDA provides a useful stylus to write in Microsoft Word, its use with large-scale documentation is tedious and difficult. This is due not only to the screen format but also to the users' limited capability to modify and extensively edit a document using a stylus. Even if new generations of mobile phones are equipped with a stylus and a larger screen more appropriate for entering information, the stylus is designed for small-scale modification demanding little written input. It is a matter of updating shorter messages rather than working with complicated documents.

IMPS's functions are built hand-in-hand with the new generation of mobile phones, which integrate more desk computer functions than before: calendar, games, picture and video functions, Internet access. From the UMS consortium's perspective, the user is a purchaser of a new mobile phone, a laptop or a PDA with real facilities for writing.

10.10 Conclusion

As a consequence of systematically comparing the industry formulation of their IMPS technology, including presence, instant messaging, groups and shared content, with a realistic portrayal of people's activities and some issues encountered when using mobile devices, we have seen the distinction between users of mobile devices and customers of new services. The comparison and evaluative ethnography reveal the confusion between users and customers within which the industry operates. From the UMS consortium's viewpoint, users have disappeared behind the mask of consumers. In the capitalistic context of the mobile phone business, Randall (2002) indicates the need for open consideration of the usefulness of technology, rather than its usability. This remark is pertinent now more than ever. We conclude with the following remarks:

1. Our adoption of an ethnographical stance in the area of the future of mobile devices shows that users of such services will get access to a wide range of electronic products. The results of those observations bring evaluative ethnography to formulate socio-economical dimensions of users rather than technical design suggestions.

2. From the UMS consortium's viewpoint, the technical integration of technology within a new mobile system is relevant to the market. Integration of the IMPS's four functions constitutes a marketing package. It is a way to sell a new generation of hardware and to attract a new generation of subscribers. Pricing of communication services is an issue that is handled within a relationship between all the partners engaged in a transaction with consumers (telecommunication network provider, hardware producers and banks). The main operators work directly with banking partners who, in exchange for providing them an incoming flux of new customers (through the billing system), offer them their financial services in return.

3. Going from CSCW studies of distributed work in large organisations to CSCW studies of the use of mobile technology in private and individualised settings allows forthcoming technology to be assessed. This study indicates a way for ethnographers in CSCW to address new technology and its development in relation to consumerism and design. It reminds any designer, who is, in the words of C. Wright Mills (1963), "a man-in-the-middle", that his or her craft depends on the kind of industrial economy he or she is involved in (a mass economy) and the way it is managed (organised in a task force team in open source communities). In the case of UMS, marketing is the structure within which technology and people operate as agents. As such, UMS's concern for users, as agents of design, is minor compared to their role as consumers of new services. People will find out what to do with a mobile phone equipped with IMPS's functions.

But to do so, they will have to buy the services first. Industrial designers seem to be the last resource of input concerning users' interest, because of their obvious stake in the industry they design for and their remoteness from people's consumption choices.

Acknowledgements

The research on which this discussion is based was supported by Stiftelsen för Internationalisering av Högre Utbildning och Forskning (STINT), the Swedish Foundation for International Cooperation in Research and Higher Education. Exchange with the faculty of Georgia Institute of Technology has helped the author to discuss aspects of this work. An earlier version of this paper was presented to the *Fourth Wireless World Conference*, 17–18 July 2003, University of Surrey, UK: The Mobile Revolution: A retrospective: Lessons on Social Shaping. I would like to thank the editors of this collection, Lynne Hamill and Amparo Lasen, for their support. I am grateful to David Hamill for his painstaking and detailed editorial work, Maria Engberg for discussions and revisions, and Dave Randall for his reflections on disappearing technology and ethnography in CSCW.

References

Arief LB, Bosio D, Gacek C, Rouncefield M (2002) Dependability issues in open source software: DIRC project activity 5 final report, CS-TR 760. Department of Computing Science, University of Newcastle. Available from: http://www.dirc.org.uk/publications/techreports/papers/10.pdf [accessed 14 October 2004].

Blomberg J, Suchman L, Trigg RH (1996) Reflections on a work-oriented design project. *Human–Computer Interaction*, 11: 237–265.

Button G (2000) The ethnography tradition and design. *Design Studies*, 21: 319–332.

Harper R (2000) Analysing work practice and the potential role of new technology at the International Monetary Fund: some remarks on the role of ethnomethodology. In: Luff P, Hindmarsh J, Heath C (eds). *Workplace Studies: Recovering Work Practice and Informing System Design*. Cambridge University Press, Cambridge; pp. 169–186.

Harper R (2002) Information that counts: a sociological view of information navigation. In: Höök K, Benyon D, Munro AJ (eds). *Designing Information Spaces: The Social Navigation Approach*. (CSCW Series) Springer, London; pp. 343–353.

Have P ten (2000) Computer-mediated chat: ways of finding chat partners. *M/C: A Journal of Media and Culture* 3(4). Available from: http://www.media-culture.org.au/0008/partners.html [accessed 10 August 2004].

Hughes J, Randall D, Shapiro D (1992) Faltering from ethnography to design. In: *Proceedings of the CSCW 1992 Conference*. Toronto. ACM, 1–4 November 1992, pp. 115–122.

Hughes J, King V, Rodden T, Andersen H (1994) Moving out of the control room: ethnography in system design. In: *Proceedings of the CSCW 1994 Conference*, Chapel Hill, ACM.

Hughes J, O'Brian J, Rodden T, Rouncefield M (2002) Ethnography, communication and support for design. In: Luff P, Hindmarsh J, Heath C (eds). *Workplace Studies: Recovering Work Practice and Informing System Design*. Cambridge University Press, Cambridge; pp. 187–214.

Mills CW (1963) Man in the middle: the designer. In: Horowitz L (ed.). *Power, Politics and People: The Collected Essays of C. Wright Mills*. Oxford University Press, London; pp. 374–386.

Murtagh G (2002) Seeing the "rules": preliminary observations of action, interaction and mobile phone use. In: Brown B, Green N, Harper R (eds). *Wireless World: Social and Interactional Aspects of the Mobile Age.* Springer Verlag, London; pp. 79–91.

O'Hara K, Perry M, Sellen A, Brown B (2002) Exploring the relationship between mobile phone and document activity during business travel. In: Brown B, Green N, Harper R (eds). *Wireless World: Social and Interactional Aspects of the Mobile Age.* Springer Verlag, London; pp. 180–194.

Orr JE (1996) *Talking About Machines: An Ethnography of a Modern Job.* Cornell University Press, Ithaca.

Palen L, Salzman M (2002) Voice-mail diary studies for naturalist data capture under mobile conditions. In: *Proceedings of the CSCW 2002 Conference.* New Orleans, ACM, 16–20 November; pp. 87–95.

Perez M (2002) Multimedia messaging services in trial phase. InstantMessagingPlanet.com, 22 July 2002. Available from: http://www.instantmessagingplanet.com/wireless/article.php/10766_1430301 [accessed 22 March 2004].

Pettersson M, Rouchy P (2002) We don't need the ambulance then: technological handling of the unexpected. In: *Proceedings of the XVth World Congress of Sociology.* Brisbane, 7–13 July.

Randall D (1994) Steps toward a partnership: ethnography and system design. In: Jirotka M, Goguen J (eds). *Requirements Engineering: Social and Technical Issues.* Academic Press, London; pp. 237–254.

Randall D (2002) Home is where the heart is. In: *Proceedings of the 3rd Wireless World Conference.* University of Surrey, Guildford, 17–19 July.

Randall D, Marr L, Rouncefield M (2001) Ethnography, ethnomethodology and interaction analysis. *Ethnographic Studies,* **6**: 31–43.

Trigg R, Blomberg J, Suchman L (1999) Moving document collections online: the evolution of a shared repository. In: Bødker S, Kyng M, Schmidt K (eds). *Proceedings of the 6th European Conference on Computer Supported Cooperative Work,* Copenhagen, 12–16 September.

Weilenmann A, Larsson C (2002) Local use and sharing of mobile phones. In: Brown B, Green N, Harper R (eds). *Wireless World: Social and Interactional Aspects of the Mobile Age.* Springer Verlag, London; pp. 92–106.

Whalen J, Vinkhuysen E (2000) Expert system in (inter)action: diagnosing document machine problems over the telephone. In: Luff P, Hindmarsh J, Heath C (eds). *Workplace Studies: Recovering Work Practice and Informing System Design.* Cambridge University Press, Cambridge; pp. 92–140.

Woods B (2002a) Ericsson deploys first wireless village IM service. InstantMessagingPlanet.com, 4 September 2002. Available from: http://www.instantmessagingplanet.com/wireless/article.php/10766_1456421 [accessed 22 March 2004].

Woods B (2002b) Wireless village releases revised specs. InstantMessagingPlanet.com, 15 July 2002. Available from: http://www.instantmessagingplanet.com/wireless/article.php/10766_1404851 [accessed 22 March 2004].

Glossary

The Open Mobile Alliance (OMA) is a consortium of interest founded by Ericsson, Motorola and Nokia. They define their initiative thus:

Wireless Village, the Mobile Instant Messaging and Presence Services (IMPS)Initiative, was formed in April 2001 to define and promote a set of universal specifications for mobile IMPS. The specifications will be used for exchanging messages and present information between mobile devices, mobile services and Internet-based instant messaging services, all fully interoperable and leveraging existing web technologies. Through its supporters, the Wireless Village initiative aims to build a vibrant community of end users and global business partners where Internet and wireless domains converge.

Available from: http://www.openmobilealliance.org/

The Institute of Electrical and Electronics Engineers, Inc. (IEEE) is a body of engineers defining itself as "An international membership organisation serving today's industries with a complete portfolio of standards programs". Available from: http://standards.ieee.org/

The Internet Engineering Task Force (IETF) is a body of engineers which defines itself as: "A large open international community of network designers, operators, vendors, and researchers concerned with the evolution of the Internet architecture and the smooth operation of the Internet. It is open to any interested individual." Available from: http://www.ietf.org/overview.html

The 3rd Generation Partnership Project (3GPP) defines:

[A] collaboration agreement that was established in December 1998. The collaboration agreement brings together a number of telecommunications standards bodies which are known as "Organizational Partners". The original scope of 3GPP was to produce globally applicable Technical Specifications and Technical Reports for a 3rd Generation Mobile System based on evolved Global System for Mobile communication (GSM) core networks and the radio access technologies that they support (i.e. Universal Terrestrial Radio Access (UTRA) both Frequency Division Duplex (FDD) and Time Division Duplex (TDD) modes). The scope was subsequently amended to include the maintenance and development of the Global System for Mobile communication (GSM) Technical Specifications and Technical Reports including evolved radio access technologies (e.g. General Packet Radio Service (GPRS) and Enhanced Data rates for GSM Evolution (EDGE)).

Available from: http://www.3gpp.org/About/about.htm

Extensible Markup Language (XML) is a language is defined by the World Wide Web Consortium (W3C; a workforce of software engineers working in groups dedicated to specific tasks) as: "a simple, very flexible text format derived from [Standard Generalized Markup Language (SGML)] (ISO 8879). Originally designed to meet the challenges of large-scale electronic publishing, XML is also playing an increasingly important role in the exchange of a wide variety of data on the Web and elsewhere." Available from: http://www.w3.org/XML/

Index

Out of print titles

Mike Sharples (Ed.)
Computer Supported Collaborative
Writing
3-540-19782-6

Dan Diaper and Colston Sanger
CSCW in Practice
3-540-19784-2

Steve Easterbrook (Ed.)
CSCW: Cooperation or Conflict?
3-540-19755-9

John H. Connolly and
Ernest A. Edmonds (Eds)
CSCW and Artificial Intelligence
3-540-19816-4

Duska Rosenberg and
Chris Hutchison (Eds)
Design Issues in CSCW
3-540-19810-5

Peter Thomas (Ed.)
CSCW Requirements and Evaluation
3-540-19963-2

Peter Lloyd and Roger Whitehead
(Eds)
Transforming Organisations Through
Groupware: Lotus Notes in Action
3-540-19961-6

John H. Connolly and
Lyn Pemberton (Eds)
Linguistic Concepts and Methods in
CSCW
3-540-19984-5

Alan Dix and Russell Beale (Eds)
Remote Cooperation
3-540-76035-0

Stefan Kirn and Gregory O'Hare (Eds)
Cooperative Knowledge Processing
3-540-19951-9

Reza Hazemi, Stephen Hailes and
Steve Wilbur (Eds)
The Digital University: Reinventing
the Academy
1-85233-003-1

Alan J. Munro, Kristina Höök and
David Benyon (Eds)
Social Navigation of Information Space
1-85233-090-2

Mary Lou Maher,
Simeon J. Simoff and
Anna Cicognani
Understanding Virtual Design Studios
1-85233-154-2

Gerold Riempp
Wide Area Workflow Management
3-540-76243-4

Elayne Coakes, Dianne Willis and
Raymond Lloyd-Jones (Eds)
The New SocioTech
1-85233-040-6

Paul Kirschner, Chad Carr and
Simon Buckingham Shum (Eds)
Visualising Argumentation
1-85233-664-1